Competition and Resource Partitioning in Temperate Ungulate Assemblies

Chapman & Hall Wildlife Ecology and Behaviour Series

Series editor: Dr. R. J. Putman, Department of Biology, University of Southampton, UK

This series brings together a number of authoritative monographs on behaviour and ecology written by experts in the field. They are concisely presented, scholarly works at research level, but written in such a way as to be readable and affordable also by the interested layman.

Each book presents an up-to-date and authoritative review of the behaviour and ecology of a particular animal species or group – but in addition, uses detailed case studies to explore a number of more general conceptual issues or principles in behavioural ecology. Developed upon the original research of the author, each account thus offers a definitive review of the species of group 'under the spotlight' with that unmistakable 'immediacy' of style of being written by someone currently directly involved with ongoing research on the animal/issues concerned; but in addition, through this detailed case-study approach, they also offer unique insights into a number of more general principles of behavioural ecology and can often provide a critical test of accepted theory.

Published

Ecology and Behaviour of the African Buffalo: Social Inequality and Decision Making
H.H.T Prins
Hardback £65.00 (0-412-44240-X), PB £24.99 (0-412-72520-7), 320pp, Nov 1995

Behavioural Ecology of Siberian and European Roe Deer
A. Danilkin (in association with *A.J.M. Hewison*)
Hardback £45.00 (0-412-63880-0), 296pp, Dec 1995

In preparation or being planned

Ecology and Behaviour of North American Black Bears: Home ranges, habitat and social organization
R.A. Powell, J. Zimmerman and D. Seaman

Comparative Behavioural Ecology of Madagascan Lemurs
J. Ganzhorn, P. Kappeler and S. O'Connor
Hardback (0-412-47250-3), Paperback (0-412-47260-0), about 256 pages

Whales, Populations and DNA
W. Amos
Hardback (0-412-44350-3), Paperback (0-412-44360-0), about 200 pages

Mole Rats, the Subterranean Foragers of Africa
N.C. Bennett and C.G. Faulkes

Fallow Deer: Behavioural plasticity in contrasting environments
J. Langbein and S. Thirgood

Behavioural Ecology of Muntjac Deer
N. Chapman and S. Harris

Evolutionary Ecology of Lemmings
Nils Chr. Stenseth

Giant Pandas of China: Ecological processes and conservation of biological diversity
K.G. Johnson

Sociecology and Conservation of Gibbons
D. Chivers

Reproductive Ecology of Phocid Seals: a North Atlantic study
W.D. Bowen and D. Boness

Competition and Resource Partitioning in Temperate Ungulate Assemblies

R.J. Putman
University of Southampton, UK

CHAPMAN & HALL
London · Weinheim · New York · Tokyo · Melbourne · Madras

Published by Chapman & Hall, 2–6 Boundary Row, London SE1 8HN, UK

Chapman & Hall, 2–6 Boundary Row, London SE1 8HN, UK

Chapman & Hall GmbH, Pappelallee 3, 69469 Weinheim, Germany

Chapman & Hall USA, 115 Fifth Avenue, New York, NY 10003, USA

Chapman & Hall Japan, ITP-Japan, Kyowa Building, 3F, 2-2-1 Hirakawacho, Chiyoda-ku, Tokyo 102, Japan

Chapman & Hall Australia, 102 Dodds Street, South Melbourne, Victoria 3205, Australia

Chapman & Hall India, R. Seshadri, 32 Second Main Road, CIT East, Madras 600 035, India

First edition 1996

Typeset in Palatino 10/12pt by Saxon Graphics Ltd, Derby

Printed in Britain by St Edmundsbury Press, Bury St Edmunds, Suffolk

ISBN 0 412 612401 2

A catalogue record for this book is available from the British Library

Library of Congress Catalog Card Number: 96-85064

Printed on acid-free text paper, manufactured in accordance with ANSI/NISO Z39.48-1992 (Permanence of Paper).

Contents

Colour plate section appears between pages 46 and 47

Series editor's foreword

Over the past 30 years or so, research effort in behaviour and ecology has progressed from simple documentation of the habits or habitats of different species to asking more searching questions about the adaptiveness of the patterns of behaviour observed; moved from documenting simply **what** occurs to trying to understand **why**. Increasingly studies of behaviour or ecology explore the function of particular responses or patterns of behaviour in individuals or populations – looking for the adaptiveness that has led to the adoption of such patterns either at a proximate level (what environmental circumstances have favoured the adoption of some particular response or strategy from within the animal's repertoire at that specific time) or at an evolutionary level (speculating upon what pressures have led to the inclusion of a particular pattern of behaviour within the repertoire in the first place).

Many common principles have been established – common to a wide diversity of animal groups, yet showing some precise relationship between a given aspect of behaviour or population dynamics and some particular ecological factor. In particular, tremendous advances have been made in understanding the foraging behaviour of animals – and the 'decision rules' by which they seek and select from the various resources on offer – and patterns of social organization and behaviour: the adaptiveness of different social structures, group sizes or reproductive tactics. Equal progress has been made in understanding the relationship of many population processes – rates of increase or mortality, relative stability of population size and structure – to environmental character.

Such principles or **explanations** of patterns of behaviour or ecology are now widely recognized and find treatment in almost any undergraduate text. Yet, important though it is to recognize such generalities of pattern, in the very derivation of such general principles often much subtlety and elegance apparent in an individual case is lost. And it may be that very subtlety which is of critical importance in fine-tuning adaptation. Indeed, it has become increasingly apparent over more recent years that there is a tremendous flexibility of behaviour and ecology even within a species. While earlier research had concentrated on exploring more general patterns, had sought to explain interspecific differences in behaviour or individual or population ecology in relation to observed environmental differences, in practice almost equal variation is apparent between different populations of one and the same species.

This new series tries to develop a focus on that subtlety of adaptation by exploring in detail the behaviour and ecology of particular species or closely related groups and trying to show how particular aspects of the behaviour or ecology described may be adaptive. In focusing on the subtlety of adaptation shown within a single species or even a single population, however, they may in addition add new perspectives, a new depth to our understanding even of those more general principles. Many of the earlier breakthroughs in developing the broader conceptual framework in behavioural ecology derived from comparative studies of closely related species – or by investigation of the wealth of increasingly recognized intraspecific variation in behaviour and ecology – and such comparative studies often reveal important new insights into the shaping of relationships between behaviour, ecology and environment.

This series then brings together a sequence of authoritative monographs on behaviour and ecology of particular species or species groups. They are concisely presented, scholarly works at research level, written by experts in their field, but also written in such a way as to be readable and accessible also by the interested layperson. Each book in the series presents an up-to-date and comprehensive review of the behaviour and ecology of some particular animal species or group, but in addition uses these detailed case studies to explore a number of more general conceptual issues or principles in behavioural ecology or population ecology.

In my own contribution to this series I have focused upon the question of how groups of species of very similar ecological requirement may coexist. Many ecological communities contain diverse guilds of species which appear to be ecologically very closely related; what factors facilitate or at least accommodate their stable coexistence? Drawing on data collected on the behaviour and ecology of just such a guild of large ungulates in the New Forest of southern England, I explore the interactions between the various species of this complex assemblage and try to understand what mechanisms control the resource relationships within the community. Lessons learnt from this particular example are then broadened to explore more generally the factors permitting and promoting coexistence in multispecies assemblages of ungulates in temperate ecosystems.

R.J. Putman
University of Southampton, UK

Preface

Although, on the basis of my published research over the past 20 years or so, most people would probably consider me primarily an applied biologist and would assess my interests as in the effects of large grazing ungulates on vegetational processes (in their impact on forestry, agriculture or conservational communities), or in the factors influencing population ecology and dynamics in terms of their implications for management, I have in truth always been intrigued by much more conceptual issues: amongst them the way in which guilds of ecologically similar species may coexist within the same community – despite an apparent potential for intense competition.

I have been extremely fortunate that I have been able over the years to foster this fascination; much of my own research work and that of my students or assistants has been undertaken within just such a multispecies system: the New Forest of southern England. Here some seven different species of large ungulates occur together – in tension or in harmony – within the 37 500 hectares of wood, heath and grassland that make up this ancient 'Royal forest'. Although very little of our early work was explicitly focused upon interactions between the different ungulate species, the data we collected over the years on the behaviour and ecology of each in turn (their patterns of resource use and their individual impact upon the Forest's vegetation) eventually became sufficiently comprehensive as to allow me to indulge my own intellectual curiosity in a retrospective analysis of how these various large ungulates do manage to coexist within the Forest ecosystem.

I published an initial exploration of such questions in 1986 (Competition and coexistence in a multispecies grazing system. *Acta Theriologica*, **31**, 271–91), but recognized that this was a very preliminary analysis of simple niche overlap based on such data as we had amassed at that time. Our work in the New Forest has continued; as the result of continuing studies of both fallow and roe deer and the Forest's populations of free-ranging ponies, we now have more detailed information for these species on patterns of habitat use and foraging behaviour than those available in 1986. In addition, other studies have provided comparable data on species which we had not previously included in any analysis, such as the Forest's small but significant population of red deer.

And while our early work within the Forest was not deliberately directed towards such analysis of interaction between the different species of large herbivore, a number of more recent studies have sought explicitly to explore those interactions. The 1986 analysis was based entirely on a synthesis of data collected for five species (cattle, ponies, fallow, roe and sika deer) in separate, individual studies; the data were not collected simultaneously for animals in the same area at the same time. To complement those 'independent' data on resource use, therefore, we have undertaken a number of studies explicitly looking at habitat use and diet of pairs or groups of species in direct sympatry.

Finally, evidence of overlap in resource use is not in itself evidence of competition. Indeed, the interpretation of measures of niche overlap in terms of the implications for competitive interaction is extremely problematical. A degree of overlap in resource use may be an essential prerequisite for competition: but observation of high levels of overlap in the field is itself ambiguous. High observed overlap **may** imply competition, but only if resources are limited; observation of high overlap might equally well be indicative of a **lack** of competition – on the basis that if severe competition were being experienced some niche-shift would have been expected, resulting in reduction of overlap. By converse, observation of low levels of overlap in the field may imply lack of competition, but may in fact reflect the end-result of changes in the ecology of some or all of the species as a direct result of competition for shared resources. In attempting to address this question we have also looked at changes in the population sizes of the different ungulates of the New Forest assemblage over the past 30 years, to see if there may be any evidence of direct interaction in the longer term – in changing abundance of one species dependent on the changing fortunes of another.

This book summarizes these new data and addresses again the question of what is going on in this multispecies assemblage. But it is not just a book about the New Forest and its large herbivores – nor yet a simple update of my earlier book on the ecology of the New Forest (*Grazing in Temperate Ecosystems: Large herbivores and the ecology of the New Forest*, Croom Helm/Chapman & Hall, 1986), a book which concentrated primarily on the effects of grazing on the Forest and its ecology. Rather it is intended as a much more general analysis of competition and resource partitioning in (largely temperate) ungulate assemblies. It may have at its core our studies in the New Forest – as a worked example running through the whole – but from that core I intend to range much more widely, drawing on a diversity of examples elsewhere in the published literature to try and derive some synthesis of the range of ecological processes that seem to be involved in structuring and maintaining such multispecies assemblies.

Where a large number of ecologically similar species seem to coexist this may be for any of a variety of different reasons. Their ecological requirements and impact may not indeed be as similar as at first suspected; thus each uses a different array of resources and interaction is slight – or nonexistent. Certain species may even actively facilitate the persistence within the community of others; through their actions and effects within the community they actually enhance the availability or quality of the resources required by some 'commensal'. But even potential competitors may coexist. If overlap in resource use is not extreme, each may restrict activity to part only of the entire spectrum of resources they could exploit in isolation, exploiting only those parts of the resource array not utilized by the competitor (demonstrating a 'shift' in the range of resources used). So long as sufficient resources remain in this uncontested 'refuge', populations of both species may still persist, supported by those parts of the resource array not utilized efficiently by the competitor. Potential competitors can also coexist within the same community if the resources shared are not limited in supply. (By definition, if there is more than enough for everyone, they cannot be competing for it.) Such a situation would not, of course, persist on its own: population sizes of one or another species would quickly expand to take advantage of the excess of resource, until resources once again became limiting. Long-term coexistence of competitors will only persist if there is frequent reversal of competitive advantage, or if some other factor keeps the population levels below that at which resources become limiting and competition becomes intense. Such factors might include the presence of predators sufficiently powerful to keep population sizes of their prey below the ceiling imposed by environmental resources, or frequent perturbation of the system, sufficiently severe and of sufficient frequency to knock expanding populations of potential competitors back down to lower levels, with such frequency that they never have an opportunity to rise to their full environmental capacity before the next decimation. There is a host of potential (theoretical) arguments which could explain the persistence of such multispecies assemblies, even the coexistence of competitors. But I am intrigued to know which of these processes operate in real-world communities; which at least are the types of interaction most prevalent among the multispecies assemblies with which I am familiar?

This book then explores the relationships and interactions between the members of a variety of temperate ungulate communities, to try to understand the processes that shape the dynamics of such multispecies guilds. The focus is deliberately on temperate rather than tropical systems because, rightly or wrongly, I feel these have been less widely reported.

Our work in the New Forest will provide a central core because it helps to provide continuity and continuity of example – and because that is the whole essence of books within this present series: that these volumes should address general issues from a case-study approach. But the work is presented in that sense: that it is a detailed, worked example from which, by relation to other examples in the literature, one may hope to draw more general conclusions.

A book such as this draws upon data and ideas amassed over the course of many years and I would like to acknowledge the debt that a synthesis such as this inevitably owes to other workers in the field. To them should go full credit for what I may attempt here; I myself accept full responsibility for any errors or misconceptions introduced in my subsequent use of those hard-won data. I would particularly acknowledge the debt I owe to colleagues and students who have worked with me within the New Forest over the years, who have guided and stimulated my own thoughts and have generously allowed me to build my current review upon their data. Particular thanks must go to Michael Boxall, Peter Edwards, Rue Ekins, Elaine Gill, Debbie Goodwin, Chris Mann, Andy Parfitt, Dave Payne, Surender Sharma and Simon Thirgood – and this book itself is dedicated to the memory of Bob Pratt, who worked with me from the very beginning and would have been amused to see where it had all ended.

I must thank the various Deputy Surveyors of the New Forest who have tolerated for so long and with such patience our work within the Forest perambulation, and the many friends we have made over the years amongst Forest keepers and Commoners; I would also thank others who have supported our efforts and have been generous with ideas and insight: I would particularly mention John Jackson, who preceded us into the Forest, Jochen Langbein and Simon Hedges. Finally, my thanks also go to Bob Carling and others at Chapman & Hall, who have helped turn this book from an idea into a reality.

Rory Putman

Technical preface

For simplicity and continuity of the text, data chapters (3, 4 and 5) have been kept as free as possible of technical detail – in order to concentrate throughout on the biological themes I am trying to develop. For the more critical reader, methods are summarized below.

Except where otherwise stated, data presented on **habitat use** are derived throughout from direct observations from fixed transects. Transect routes were established in each study/study area to traverse all different vegetation types present, sampling each in approximate proportion to their abundance within the site as a whole – and, where possible, sampling a number of different patches of each such habitat. Transect routes were between 2 and 5 km and were walked or driven during daylight hours for all species, on a minimum of three different occasions (at dawn, midday and dusk; fallow deer: Thirgood 1995a), 4–6 (ponies: Gill 1988; roe deer: Sharma 1994) to a maximum of once every 2 hours (cattle and ponies: Pratt *et al.* 1986; sika deer: Mann 1983, Mann and Putman 1989a); in studies of sika deer, cattle and ponies only, routes were also traversed during the hours of darkness – on three occasions each sampled night (sika deer: Mann 1983, Mann and Putman 1989a) or at 2-hourly intervals (cattle and ponies: Pratt *et al.* 1986). Transects were repeated on a number of separate sampling 'days' each month over a 12-month period.

Of necessity, survey transects could not sample all vegetation types in perfect proportional relationship to their availability within a site. Estimates of habitat usage within the site as a whole were therefore made by weighting the numbers of individual animals observed (O_i) in any habitat (i) by the relative area of that habitat surveyed (S_i) in relation to its total availability within the site (T_i). Proportional use of any habitat (U_i) is thus presented as the proportion of total observations after weighting, as:

$$U_i = [O_i/(S_i/I_i)]/\Sigma[O_i/(S_i/T_i)].$$

Visibility on transect routes was not constant throughout the year, due to leaf fall or growth of the understorey; for this reason, separate estimates of the area of each habitat surveyed (S_i) by any transect route were made in each of the four seasons of the year.

Where proportional use made of different habitats is examined in relation to relative availability of those habitats within the animals' range, indices of **habitat selection** calculated are modifications of Hunter's index (Hunter 1962) by Pratt *et al.* (1986) or Goodall (in Hirst 1975).

Dietary analyses for fallow and roe deer presented by Jackson (1977, 1980) are based on analyses of ruminal contents, following the methods of Chamrad and Box (1964). In all other cases (unless explicitly stated in the text) dietary composition reported is based on microhistological analysis of faecal materials. Analyses followed standard protocols (after Stewart 1967, Stewart and Stewart 1970, Putman 1984).

Samples of fresh faecal material were collected in the field during the course of other routine observations. For cattle and ponies a minimum of 15 individual boluses were collected in each month (Putman *et al.* 1987; Gill 1988); for the deer species, 15 pellets or more from each of 15 defaecations. All samples were air- and then oven-dried for storage before analysis. In analysis, a number of pellets from each pellet group were mixed and this bulked sample was ground with a pestle and mortar, using a solution of 2% NaOH or KOH as a lubricant, and left soaking for 12 hours. The supernatant was poured off and replaced with 10% NaOH/KOH and the sample boiled for 5 minutes in a fume cupboard. This treatment ensures separation of all the particles and also removes gut mucus and bacteria, rendering cuticular fragments more translucent. After cooling, the sludge was washed and centrifuged to compact the faecal matter. A portion of the cleaned material was then spread, in water, on a glass plate or Petri dish engraved with a square grid before examination under a binocular microscope at 90× magnification. The first 100 fragments lying across or touching grid lines or intersections were identified by comparison with a standard set of reference slides/photomicrographs, with the same reference collection common to all studies. Three replicate dishes of each sample were scored; monthly dietary profiles were assessed as the number of fragments of each identified forage species in three plates of (minimum) 15 samples, expressed as a proportion of the total number of fragments scored.

Statistical analyses of patterns of habitat use or dietary composition were mostly made, in the original publication, using non-parametric methods: primarily employing a G-test for homogeneity on initial frequencies (Sokal and Rohlf 1981); in most cases these data have subsequently been re-analysed here using multivariate analysis of variance (MANOVA). Plant fragments recorded in any faecal sample, or numbers of animals observed in any habitat, strictly represent quantitative numerical values rather than true frequencies; however, with a finite number of fragments scored for any sample – or a finite number of animals to be dispersed among possible habitat classes along a transect route – values recorded in each category are not statistically independent.

Data were thus transformed using the additive logistic transformation of Aitchison (1982), quoted by Wilkinson (1986), in which $n+1$ original variables, x_i, are converted into n new variables, y_i, as $y_i=\ln(x_i/x_{i+1})$. In effect all values for fragment number in any dietary category are divided through by the number in the most frequently recorded category and then logged; similarly, the weighted number of animal observations attributed to each habitat is divided throughout by the number of observations in the habitat most used – and those ratios logged. The new variable set contains all the information of the old, and each new variable is now approximately normally distributed and independent of the others.

In statistical analysis by MANOVA, most analyses employed a one-way design, using different forage categories or habitat types as multiple dependent variables to test the effects of a single categorical variable at a time (sex, month, season) on differences in the proportional use of habitat/dietary components. Where attention is drawn in the text to differences in patterns of habitat use or in dietary composition between months/seasons, between sites, between sexes or between species, such differences are statistically significant at $P < 0.05$ – and usually with $P < 0.01$ or better.

The majority of the studies quoted here have already been published in the primary literature. Readers requiring still more detail of methods used are advised to refer back to those original publications:

Habitat use:
cattle: Pratt *et al.* (1986)
ponies: Pratt *et al.* (1986), Gill (1988)
sika deer: Mann (1983), Mann and Putman (1989a)
fallow deer: [Jackson 1974], Thirgood (1995a)
roe deer: Sharma (1994).

Diet:
cattle: Putman *et al.* (1987)
ponies: Putman *et al.* (1987), Gill (1988)
sika deer: Mann (1983), Mann and Putman (1989b)
fallow deer: [Jackson 1977], Putman *et al.* (1993)
roe deer: [Jackson 1980], Sharma 1994.

1

Introduction

1.1 MULTISPECIES SYSTEMS AND THE POTENTIAL FOR INTERACTION

In almost any terrestrial ecosystem – temperate or tropical, New World or Old World – it is common experience that herbivores outnumber carnivores. At the most simplistic level one would expect such reduction of abundance (numbers or biomass) as one ascends a simple trophic pyramid; but it is equally apparent that not only are the numbers or biomass of herbivorous animals far greater than those of the carnivores: there is also a far greater diversity of species. Certainly this is strikingly evident among assemblages of larger mammals; we are accustomed to high diversity in the tropics, but even in temperate regions, woodland and grassland systems are characterized by a surprising diversity of larger herbivores. Historically, such large herbivore assemblies were perhaps even more diverse than they are today, but even now, despite the historical intervention of man in the elimination of many of the larger species, many terrestrial systems support rich, multispecies assemblages.

Yet in many ways, this presents something of an anomaly. How do these multispecies systems persist without conflict? What mechanisms facilitate the stable coexistence of so many similar species? A simple answer might be that they are all ecologically so distinct that they simply do not interact – or that resources are so abundant that even similar species are never brought into competition; but neither solution bears close inspection.

Many of the herbivore species within any given assembly clearly have very similar feeding styles and foraging habit, grouped into clear functional guilds of grazers, concentrate selectors or intermediate feeders (mixed browser/grazers) (Hofmann 1973, 1985); many, particularly amongst the grazers, are relatively unselective foragers, with diet composed simply of those forages in greatest abundance; diets clearly, commonly show extensive overlap. However, herbivores in general remove less than 10% of the above-ground primary production from **any** natural community (more commonly nearer 5%), and large herbivores on their own – as distinct from smaller rodent or invertebrate herbivores – generally consume far less than this (Petrusewicz 1967). We might thus presume that resources are superabundant, permitting a great diversity of species to coexist in harmony since, with no shortage of resources, they will experience no effect of competition from others around them. But this answer is also far from satisfactory.

If resources were genuinely superabundant, one might expect the numbers of individuals in one or more of the different herbivore species to increase to take advantage of these 'surplus' resources, until the spare capacity had been taken up. Superabundance of resources would not persist indefinitely – unless numbers of each species were regulated below resource capacity by some other factor. Indeed, it seems probable that this statistic, that large herbivores typically consume something less than 10% of above-ground primary production, may actually be somewhat misleading anyway: it does not automatically follow that the remaining 90% is in fact available to them. Recent analysis of plant (physical and) chemical defences certainly suggests that a large proportion of nutrients apparently represented by this material may not be accessible by herbivores at all, and that in many systems herbivores are genuinely resource limited. All of which suggests that our multispecies guilds do not persist by escaping interaction.

1.2 MECHANISMS OF COEXISTENCE

In theory, there are several potential mechanisms whereby species of similar ecologies may coexist within the same community (and see Putman 1994, for a review). Thus, as we have already suggested, they may coexist because, in effect, their characteristic patterns of resource use are from the outset distinct and non-overlapping. More formally, we might say that there is in effect some separation in fundamental niche between the species – not a separation achieved by change of resource use in the presence of each other, but a fundamental difference even in resources preferred in the absence of interaction. In such case (and one might consider a preferential grazer such as a fallow deer alongside an obligate browser/concentrate selector such as a roe deer), there is no potential for interaction even in coexistence. The example given, however, does not stand close scrutiny (the separation suggested in this case is by no means as complete as presented!) and indeed it is rare to find situations where there is absolutely no overlap in fundamental niche.

However, overlap in patterns of resource use need not necessarily be detrimental. There is potential in such cases for the interaction to be mutualistic – or at least of no disadvantage to one partner in the interaction while positively beneficial to the other. Such **facilitation** has been suggested in the grazing succession of ungulates in the Serengeti National Park (Gwynne and Bell 1968, Bell 1970, Sinclair and Norton Griffiths 1982).

Here, within the wider array of larger herbivores feeding on the short-grass plains of the Serengeti, four species of ungulates apparently all feed on the same species of grass without any signs of competing. The animals (zebra, *Equus burchelli*; topi, *Damaliscus korrigum*; wildebeest, *Connochaetes taurinus*; and Thomson's gazelle, *Gazella thomsoni*) occupy the area in strict sequence. First the zebra eat the upper parts of the stem; these are low in protein, but as a hind-gut fermenter, the zebra's digestive system is adapted to deal with low quality feed by having a high rate of throughput. Then the topi eat the lower stem (which contains more protein than the upper part). The wildebeest eats the leaves, while the Thomson's gazelle, following behind, eat the new green shoots which spring up within a day or two of the plant being cropped back by the other animals. Thomson's gazelle also eat the dicotyledonous plants growing at ground level, reducing the forb canopy and thus permitting regeneration of the annual grasses, replenishing the sward for the other ungulates in the guild. In effect it has been argued that each species 'prepares' the sward for the next in the succession.

Although this grazing succession in the Serengeti offers elegant demonstration of the principle of feeding facilitation, more recent analysis urges caution in interpreting data based simply on association (de Boer and Prins 1990), pointing out that in this particular example the conclusion that one species benefits from the earlier grazing by another is not proven – and indeed such assumption is not supported by data on the respective population dynamics of the species concerned (Sinclair and Norton-Griffiths 1982, Stelfox *et al.* 1985). However, such criticism does not invalidate the principle of facilitation – and in a more convincing, temperate example, grazing by cattle was shown to increase quality of grass swards available to red deer hinds on the Isle of Rum (Gordon 1988).

Even where species show substantial overlap in fundamental niche with clear potential for competitive interaction there are several ways in which stable, or at least persistent, coexistence may be achieved. Coexistence of potential competitors may result from a change of resource use in interaction. As long as overlap in potential resource use (fundamental niche) is incomplete, and there remains for each species a part of the resource array which they alone may exploit, a shift in the overall pattern of resource use may be observed in response to the presence of the other species, so that one or other species 'withdraws' from the zone of overlap, concentrating instead on those parts of the resource not utilized by other species. In effect while not so different that they differ in fundamental niche (above), they none the less achieve separation of ecology in practice, as each comes to occupy a distinct and non-overlapping 'realized' or 'post-interactive' niche (terms after Hutchinson 1957, Vandermeer 1972) when in the presence of the other(s). Yet other mechanisms permit the coexistence of potential competitors even without resolution of overlap in resource use.

Thus, as we have already noted, coexistence of potential competitors may be accommodated where resources are in superabundant supply; by simple definition, if there are more than enough resources for all, they cannot suffer by having to share those same resources with others. While we have suggested a superabundance of resources would be unlikely to persist for long in the absence of other factors – in that one or another species might be expected to increase in abundance to 'take up the slack' – populations of even fierce competitors might persist in stable coexistence if population densities were kept below the level at which resources become limiting by some other agency. Where, for example, the effects of predation (by a single generalist predator or a suite of species each specializing on one or more within our guild of herbivores) act to reduce populations of all potential competitors below the level at which they would be limited by resource availability, all may coexist. (In theory such coexistence of competitors could also be accommodated where competition is uneven, if a single specialist predator acted in such a way as to reduce the vigour of the most powerful competitor within the guild.) Other factors too may facilitate coexistence of competitors by keeping populations below levels at which resources actually become limiting.

Huston (1979) has shown that the introduction of occasional episodes of heavy density-independent mortality into classical Lotka–Volterra models of competition can delay almost indefinitely the final resolution of exclusion; in effect in any community subject to regular disturbance of a severity sufficient to disrupt population growth, potential competitors may persist without exclusion simply because population sizes of none of the species concerned ever get a chance to reach a level at which resources would become limiting, before the community's dynamics are once more disrupted by the next perturbation (see for example Connell 1978, Miller 1982, Chesson and Case 1986). This realization that coexistence of competitors within a community may result from fluctuations in environmental conditions of such a periodicity that competitive interactions are prevented from running to conclusion has led to the recognition of a number of other non-equilibrium models of competitive coexistence (see for example Putman 1994). But all, in practice, envisage coexistence as permitted because populations of the potential competitors are maintained below a level at which resource limitation would become apparent by external agencies – whether biotic (predation) or abiotic (climatic fluctuations, fire).

There is of course one further possibility we have not yet considered: that at least some members of a multispecies complex are in active competition, that such assemblages may not persist indefinitely, but that given time one or another species may ultimately be excluded.

1.3 THE EVIDENCE FOR COMPETITION

That there genuinely is a real potential for such competitive interaction is beautifully illustrated by a unique set of enclosure experiments in Texas (Harmel 1980, 1992) investigating the degree of competition for native white-tailed deer (*Odocoileus virginianus*) posed by three introduced exotics: fallow deer (*Dama dama*), chital (*Axis axis*) and sika deer (*Cervus nippon*). In each case six adult white-tailed deer were introduced, together with six adults of one of the other species, into a 39-hectare enclosed area of native vegetation in the Edwards Plateau region of Texas; population number of each species was monitored within the enclosure from introduction for a period of 6 years (fallow) to 8 years (chital, sika). A similar 39-hectare enclosure, operated since the early 1950s with about 15 white-tails, was used as a control.

The control population remained stable throughout the period of the trials – at a mean number of 14 individuals (or about 0.38 deer per hectare: around the average density for the region as a whole). In the mixed sika/white-tail enclosure, sika deer numbers increased from 6 to 62 from introduction in 1971 to the end of the experiment in 1979 (Figure 1.1a); the population of white-tails increased to 17 in the second year – and then declined to extinction. In the chital enclosure, population numbers of white-tailed deer again increased initially, to a peak of 11 individuals in 1975, before declining again to a total of three non-breeding animals by 1979; over the same period numbers of chital rose from six to stabilize between 15 and 20 individuals (Figure 1.1b). Enclosure studies of fallow deer with white-tails were not initiated until 1986; in this case, white-tailed deer populations persisted in the enclosure at densities similar to those in the control pen while fallow deer numbers had declined to zero by 1992, 6 years after the commencement of the trial. In all three cases mixed stocking within the enclosures resulted in a suppression of performance of one of the species – sufficient over time to lead to its elimination from the 'community' – while the other species persisted at densities equivalent to those observed for white-tailed deer in the control pens or outside rangelands: results consistent with direct competition between the species concerned.

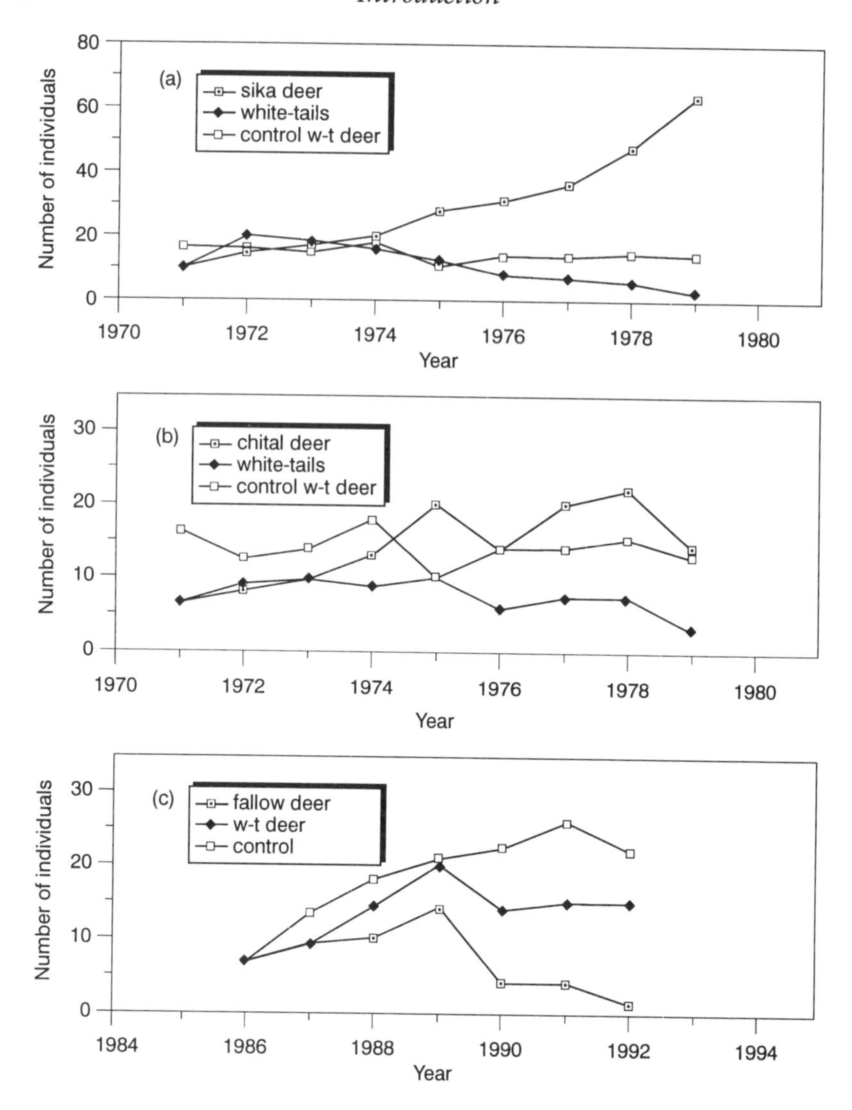

Figure 1.1 Population trends of white-tailed deer and of sika, chital and fallow deer in mixed species enclosures, see text for details. Data from Harmel (1980, 1992) show: (a) when sika deer were introduced into enclosures with white-tailed deer, sika populations expanded while white-tailed deer populations dwindled to extinction; (b) in pens where chital deer were enclosed with white-tails, white-tailed deer populations again declined to extinction. However, (c) white-tailed deer maintained healthy populations when in mixed stocking with fallow deer; in this case the fallow deer were eliminated. (Reproduced from Feldhamer and Armstrong 1993, with permission.)

The real world is, however, more complicated. In an experimental situation two species can be brought into explicit confrontation over a rather restricted array of resources, prevented by the restrictions of the experiment itself from exercising the degree of partitioning that might be observed in a more complex environment. Species that overlap considerably in their use of one resource may coexist without interaction if sufficiently different in their use of some other orthogonal resource (Pianka 1976, 1981); similarly, heterogeneity of environment in both space and time, leading to repeated reversal of the competitive advantage of one species over another in different temporal or spatial 'patches', may facilitate prolonged coexistence of species which would readily compete to exclusion in a more homogeneous environment. (An excellent empirical illustration that spatial heterogeneity does indeed facilitate the coexistence of competitors which compete to extinction in simple laboratory 'bottle experiments' is provided, for example, in the work of Atkinson and Shorrocks on *Drosophila* (1981, 1984) while the role of temporal variation in conditions is demonstrated elegantly in Spiller's (1984) analysis of seasonal reversals in competitive advantage between two orb-weaving spiders, *Metepeira grinnelli* and *Cyclosa turbinata*.)

Amongst large ungulates, evidence for interaction in the natural state is limited. Studies of such interaction in completely natural systems are relatively scarce and where such systems have been explored, while the separate ecologies of individual species have been documented, the mechanism and scale of actual interaction between the species have often not been considered explicitly. Most commonly, such studies consider simply the degree of overlap observed in patterns of resource use.

Mathur (1991), for example, discusses ecological separation between sympatric populations of chital, sambar and nilghai in three National Parks in India, in relation to use of both habitat and forage resources; in earlier studies in Nepal, clear separation in the use of habitat was recorded between chital, hog deer and barasingha (Schaaf 1978) and between chital, sambar, hog deer and muntjac in Chittawan National Park (Mishra 1982). The individual habitat preferences of sable antelope, waterbuck and impala in southern Africa are presented in an elegant analysis by Ben-Shahar and Skinner (1988); resource relationships of ungulates of the intensively studied Serengeti ecosystem are summarized by Jarman and Sinclair (1979). In none of these cases are the mechanisms or implications of such partitioning explored in detail.

In temperate systems too, the majority of studies merely **describe** patterns of resource use and partitioning among sympatric species: (e.g. Hudson 1976, for white-tailed deer, mule deer, wapiti and bighorn sheep in British Columbia; Cairns and Telfer 1980, for moose, wapiti, white-tailed deer and American bison; Jenkins and Wright 1988, for white-tailed deer, wapiti and moose in Montana; Grodzinski 1975, Jedrzejewska *et al.*

1996, summarizing work of various authors on the ecology of roe deer, red deer, fallow deer, moose and European bison in Polish forests; Chapman *et al.* 1985, for ecological separation between roe deer, fallow deer and Chinese muntjac sympatric within a commercial conifer forest in the UK; Homolka 1993, for dietary overlap between red deer, roe deer and moufflon in the Czech republic). What we cannot tell from any of these studies is whether or not the clear ecological separation expressed implies a lack of any potential for interaction, or whether that separation is itself an explicit response to competition.

Indeed, the interpretation of measures of niche overlap in terms of the implications for competitive interaction is extremely problematical. A degree of overlap in resource use may be an essential prerequisite for competition: but observation of high levels of overlap or separation in the field is itself ambiguous. High observed overlap **may** imply competition, but only if resources are limited; observation of high overlap might equally well be considered indicative of a **lack** of competition – on the basis that if severe competition were being experienced some niche-shift would have been expected, resulting in reduction of overlap. By that same token, observation of low levels of overlap in the field may imply lack of competition – but may in fact reflect the end-result of changes in the ecology of some or all of the species as a direct result of competition for shared resources. Only where we may find evidence of a clear shift in resource use of a species in allopatry and sympatry may we suspect a competitive interaction – or better still when clear overlap in resource use is accompanied by an inverse relationship in population sizes of a given species pair.

In Kanha National Park in central India for example, where a clear dietary overlap has been recorded between chital deer and barasingha in their use of open grassland species, a recent decline in population numbers of barasingha has coincided with a large increase in abundance within the park of chital (Martin 1987). Chapman *et al.* (1985) recorded high overlap in diet and winter habitat use of roe deer and Chinese muntjac in the King's Forest in Suffolk, UK (Chapman *et al.* 1985, Forde 1989) and there is increasing evidence of a decline in numbers of roe in areas of high muntjac density (Forde 1989, Wray 1994).

By contrast, detailed analysis of population trends of roe deer, red deer, fallow and sika in the New Forest of southern England over a 24-year period (Putman and Sharma 1987) failed to show any consistent correlation between population sizes of the four species, despite significant apparent overlap in resource use (Putman 1986a). In the absence of experimental manipulations like those of Harmel (1980, 1992), however, even this type of evidence is little more than circumstantial – based essentially on simple correlations with no proof of any causal relationship – and, in practice, investigation of competition in the field is extremely difficult.

1.4 ESTABLISHING COMPETITION IN NATURAL COMMUNITIES

Recognition of the difficulty of establishing competition in the field – amongst any group of organisms – has led various workers to try to develop formal protocols for adducing competitive interaction in natural communities. Wiens (1989) has modified and extended the criteria developed earlier by Reynoldson and Bellamy (1970), suggesting a range of types of evidence which may be sought of different levels of cogency. Thus they argue that although competition can rarely be proven without doubt, ever-increasing strength of supposition may be drawn from:

Weak:
1. Observed patterns (of population trend or shifts in resource use between sympatry and allopatry) are consistent with predictions from competition.
2. Species overlap in resource use.
3. Intraspecific competition occurs.

Suggestive:
4. Resource use by one species reduces availability of resources for another species.
5. One or more species is negatively affected.

Convincing:
6. Alternative process hypotheses are not consistent with observed patterns.

Wiens' criteria are stringent – and satisfaction of all is rarely practicable in the field. An alternative logic is proposed by de Boer and Prins (1990). In a sense, de Boer and Prins argue by converse: seeking not to prove competition, but rather (on the basis that it is often easier to disprove a hypothesis rather than prove it beyond doubt) establishing hurdles of definite disproof. Investigating the potential for competition between African elephant (*Loxodonta africana*), Cape buffalo (*Synceros caffer*), wildebeest and Burchell's zebra within the Lake Manyara National Park of Tanzania, they argue that interspecific competition between any two species is only possible where three separate conditions are met (de Boer and Prins 1990):

1. There must be habitat overlap.
2. There must be overlap in forage consumed by the two species within those shared.
3. The shared dietary resources must be limiting.

In this book I intend to re-examine the available literature and try to resolve further the patterns of interaction that may be observed in multi-species assemblies of large ungulates, to attempt to unravel something of the complexity of interaction. We will be seeking for evidence of facilitation or competition, and, in investigating the extent of competition observed, applying these more stringent criteria of Wiens (1989) and de Boer and Prins (1990). Ultimately I seek to resolve two main questions: just how far the different species within such assemblies do influence each other's ecology and population dynamics, and how, even in modern times, quite species-rich assemblages of closely related species manage to coexist within a single community. I will focus in particular on temperate systems, since, rightly or wrongly, I believe these to have been less reported, and for a major part will develop the analysis through detailed examination of one particular case study. Already in this introduction I have illustrated one point or another by reference to the New Forest in southern England. I intend to devote the next few chapters to a detailed examination of relationships and interactions between the seven large herbivores which share this complex system, before returning in a final chapter to place the conclusions that may be drawn from this worked example into wider context.

2

The New Forest and its larger herbivores

2.1 THE NEW FOREST

Perhaps only the English with their unwitting natural irony could have retained the name New Forest for an area that is neither new (it is one of the oldest semi-natural areas of woodland in Great Britain, as well as one of the largest) nor what most would regard as a forest. 'New' only when it was first created the latest in a series of 'Royal Forests' in the eleventh century, it is also a 'forest' only in the medieval sense of an area set aside as a royal hunting preserve (and thus coming under Forest rather than Common law). Set in the Hampshire Basin of southern England, between the Solent and the Avon (Figure 2.1) this 'New Forest' comprises a diverse mix of vegetational communities: only some 10 000 ha (of a total administrative area at the current time of approximately 37 500 ha) are actually forested in the sense of covered with trees; the remaining area is a complex mosaic of open heathland, grasslands and bog.

The diversity of vegetation types within the Forest is in fact far greater than might be expected for the area and this may be attributed to considerable variation in edaphic factors over the area. Heavily leached and base-poor plateau gravels are widespread, particularly to the north, and support a *Calluna*-dominated **dry-heath** community. At lower altitude, and where the plateau gravel has been eroded, more fertile clays and loams support mixed **deciduous woodland**. This is predominantly of beech and oak, with an understorey of holly; common bent (*Agrostis capillaris*) colonizes the woodland floors in openings and glades. (Many of the more fertile woodland sites have been enclosed over the past 100 years and now support commercial plantations: largely coniferous.) Also common on these more fertile soils is a range of natural **acid-grassland** communities, dominated by the coarse *Agrostis curtisii* and to a lesser extent by the purple moor-grass (*Molinia caerulea*), usually colonized to a greater or lesser extent by bracken (*Pteridium aquilinum*) and also by extensive brakes of gorse (*Ulex europaeus*).

Figure 2.1 Location of the New Forest in the Hampshire Basin of southern England.

Where drainage is impeded on the lower slopes, domination of the heathland community by *Calluna vulgaris* is diminished and the species diversity of the whole heathland increases. A clear gradation is observed from the dry-heath association through **humid** and **wet heath**, with increasing abundance of cross-leaved heath (*Erica tetralix*) and *Molinia* and the appearance of true wetland plants such as bog asphodel

(*Narthecium ossifragum*) and *Juncus* species. This progression frequently ends in the development of a bog community. These **valley bogs** offer some of the richest communities in the New Forest in terms of plant diversity and are one of the formations unique to this area. The species composition varies considerably in relation to the degree of eutrophication, and several distinct communities may be recognized. Perhaps the most widespread in base-poor water is that dominated by tussocks of *Molinia,* with common cottongrass (*Eriophorum angustifolium*) and *Sphagnum* mosses abundant between the tussocks. In many heathland catchments, **carr woodland** communities develop in the valley bottoms where drainage waters have a definite axis of flow. These carrs are composed of *Salix atrocinerea,* alder buckthorn (*Frangula alnus*), alder (*Alnus glutinosus*) and other tree species, and have a diverse herb layer including the greater tussock-sedge (*Carex paniculata*)

Where the area is well-drained by one of the many small streams that dissect the Forest, the bogs are replaced and the heathland progression terminates abruptly at the edge of alluvial strips bordering the streams. These alluvial deposits are covered by grassland, often dominated by *Agrostis canina,* interrupted with patches of riverine woodland. These **streamside lawns** are particularly nutrient-rich because of regular annual flooding from the rivers they border, which carry base-rich compounds from north of the Forest area.

Very little of the New Forest vegetation can be considered entirely natural and most areas have at various times been subjected to human management. Heathland communities, for example, created originally by extensive livestock grazing of Saxon times, nowadays are maintained in a programme of regular cutting or controlled burning, so that any extensive area of heath contains a patchwork of subcommunities from 0 to 12 years of maturity. Woodlands, even ancient deciduous blocks, are commonly of artificial structure and origin, planted initially for timber production, even if subsequently left more to processes of natural regeneration and decay; the 8000-odd hectares of commercial coniferous forest established in the past 100 years are clearly also of entirely artificial origin. In addition, there are other distinct community-types of anthropogenic origin.

A number of areas of the natural acid grassland of the Forest were fenced during the Second World War, ploughed, fertilized and cropped for potatoes or oats. At the end of the war, these areas were reseeded with a commercial ley, and after the grassland had become established, the fences were removed to return these **reseeded areas** to the Forest grazing. In the late 1960s and early 1970s a number of other attempts were made to improve the Forest grazing, by swiping bracken from other areas of acid grassland and liming them. These **improved areas** once again form a distinct and characteristic vegetational community.

All these different vegetational types are patchworked together into an intimate mosaic throughout the Forest. Of the total administrative area of 37 500 ha, some 9000 ha are in fact occupied by towns, villages or agricultural land; 8300 ha have been enclosed for commercial forestry; while some 20 000 ha remain as 'wasteland' or 'Open Forest', whose vegetation we have just described. This 20 000 ha, with its complex mix of vegetational communities, supports a diverse and numerous assemblage of large herbivores. Some 2500 wild deer (red, roe, sika and fallow) have access to the entire area, while the open unenclosed Forest supports in addition large numbers of ponies and cattle, pastured on the Forest grazings under ancient Rights of Common.

Of the four main deer species present on the Forest today, red deer (*Cervus elaphus*) and sika deer (*Cervus nippon*) populations are of relatively recent origin and are essentially local in distribution; populations are restricted to relatively limited areas of the Forest as a whole, although both are currently expanding their range. Roe deer (*Capreolus capreolus*) are distributed more widely, but the distribution is patchy and they are everywhere uncommon. Fallow deer (*Dama dama*) are both widespread and abundant, with numbers estimated in recent years at around 2000 (Strange 1976, Putman and Sharma 1987). Reeves' muntjac (*Muntiacus reevesi*) have also recently become established in the Forest as residents, but numbers are low and as yet they have had no pronounced impact upon the Forest community.

Domestic animals are also grazed upon the 20 000 ha of the Open Forest. One of the concessions granted to the local populace after the declaration of the area as a Royal Hunting Forest was the right of Common Grazing. On the payment of an appropriate 'marking fee' local cottagers and farmers could turn out cattle and horses to exploit the rough grazing of the Forest lands. These rights are still honoured and large numbers of cattle and ponies are regularly pastured at free range upon the Forest grazings. Finally, ancient rights of 'pannage' also provide for the turning out of pigs into the Forest's woodlands for a restricted period in the autumn, to feed upon the rich crop of tree-fruit: acorns and beechmast.

The numbers and relative importance of all these herbivores have fluctuated over the years. In times past, the Forest deer were probably by far the most significant grazing pressure upon the New Forest vegetation. The area was, after all, set aside primarily for the preservation of game, and until the 1850s deer populations over the area numbered between 8000 and 9000 head. (Throughout this time, the deer would have been predominantly fallow, with at most a few hundred red deer.) At the end of the nineteenth century, however, the area was 'disafforested'; deer populations were decimated and have only recently recovered to their present levels. Their present, reduced abundance now casts the free-ranging domestic livestock as the Forest's major herbivores. Cattle and ponies have been depastured on the Forest alongside the deer ever since its designation as a Royal Forest and probably considerably before that time, but numbers were probably much lower than at present (each commoner probably only turned out one or two ponies and cattle, sufficient to support an essentially cottage economy) – and in addition a far larger area of land was unenclosed and available for common grazing. As a result, the impact of the common stock was probably secondary to that of the deer. However, with the reduction of numbers of deer and simultaneous increased effective density of domestic stock at the end of the nineteenth century, cattle and ponies emerged as the major grazing influence and have remained so to this day.

But whether by deer or domestic stock, the New Forest area has always sustained a tremendous grazing pressure from large herbivores. At present, 20 000 ha of some of the poorest possible grazing (current land-use survey maps class the majority of the area as grade 5, or non-agricultural land) support a total large herbivore biomass in excess of 2500 tonnes, and it is clear that equivalent grazing pressure must have existed over the centuries. This history of continued grazing has stamped its mark on the Forest vegetation – and indeed the current ecology of the entire area can only be interpreted in relation to the various effects of past and present grazing. Any attempt to describe the ecology of the Forest -- in accounting for the curious lack of diversity of many of the vegetational systems, the low numbers and diversity of small mammals, curious behaviour of birds of prey and other predators – any attempt to explain rather than simply describe, forces the attention back to the dominating effect of grazing in the shaping of this ecosystem. Even my descriptions above of the major communities of the Forest are bedevilled by the fact that they themselves, in physical architecture and species composition, are shaped as they are by centuries of grazing. These effects of grazing past and present on the ecology of the Forest formed the focus of an earlier book (Putman 1986b) and do not need to be rehearsed in any great detail here; so pervasive is their influence on the Forest and its entire ecological functioning, however, that they are bound to intrude into later chapters and it may be helpful at least to summarize briefly some of the major implications at this point. (Such treatment in this context is necessarily brief; for fuller discussion the reader is referred to Putman 1986b, 1987, Tubbs 1986.)

2.2 THE EFFECTS OF GRAZING IN THE NEW FOREST

Grazing and browsing from wild ungulates have always played a major role in determining the structure and dynamics of natural ecological systems. Indeed, both in terms of their immediate, present-day influence on composition and function and as a powerful selection pressure in the original development of such systems and their species characteristics, large herbivores may commonly constitute one of the most significant forces shaping terrestrial ecosystems.

The effects of grazers and grazing upon the vegetation are far-reaching: grazing may directly affect species composition, diversity, productivity, even the physical architecture of the plant community. Patterns of foraging and elimination may affect nutrient dynamics and patterns of nutrient flow, with further implications for plant species composition, distribution and productivity.

Further, effects of grazing are not limited to an influence on vegetational structure and dynamics. Through their impact on the composition and productivity of the vegetation these herbivores immediately have a secondary and equally significant influence upon all other animals dependent on that same shared vegetation, affecting in short the composition and dynamics of the entire community.

Almost all these effects may be registered within the various vegetational communities of the New Forest. Clear changes in species composition in response to grazing, with selective elimination of species particularly sensitive to defoliation, or others more tolerant but heavily preferred, are apparent in most of the Forest communities accompanying a gross shift in community structure towards those species which are in some way more resistant to, or tolerant of grazing impact.

Such changes over time have been recorded already in the brief 30-year history of grassland areas ploughed and reseeded after the Second World War and only opened to the Common grazing in the early 1960s (Pickering 1968, Putman *et al.* 1981, Putman 1986b), with the establishment at equilibrium of communities dominated by stoloniferous grasses (such as *Agrostis capillaris*) and prostrate or rosette-forming herbs such as daisies (*Bellis perennis*), catsear (*Hypochaeris radicata*) or plantain (*Plantago lanceolata*), which by their growth form are more able to withstand or escape grazing.

Species composition and horizontal patterns of distribution within these same grassland communities are also affected by clear 'dislocation' of nutrient return, whereby feeding patterns of the various large herbivores result in a very patchy and discontinuous return to the system of abstracted nutrients. Animals that forage over a relatively wide area but defaecate in a smaller area can have a substantial impact on local nutrient availability. Sheep, for example, graze widely over a pasture during daylight hours but congregate in camps at night or for shade; in consequence, 35% of their faeces are deposited on less than 5% of the grazing area – resulting in a gradual impoverishment of the wider grazing range but continued enrichment of small areas within it (Spedding 1971); these patterns of grazing and elimination result in the development of a fine-scale heterogeneity of species associations within swards grazed by sheep (Bakker *et al.* 1983a,b). Such 'nutrient dislocation' has also been recorded for horses (Archer 1973, Edwards and Hollis 1982), which establish distinct and fixed grazing and latrine areas in different parts of their foraging range. Edwards and Hollis (1982) showed that free-ranging ponies of the New Forest, like their more domesticated counterparts in fields, established within their grazing grounds distinct and traditional sites for grazing and for elimination. The animals cropped swards close in areas selected for grazing – and undertook specific and purposeful movements away from these areas to defaecate and urinate in traditional latrine sites, within which they did not graze except in occasional periods when other forage was extremely scarce. These traditional latrine sites were fixed and persisted in the same areas for year after year, establishing a clearly non-random pattern of return of nutrients within the community, which was not masked or reversed by the activities of other grazers. Although cattle and deer also utilized these Forest grasslands, their feeding was restricted to the pony latrines. With incisors in both upper and lower jaw, the ponies can crop the sward in grazing areas so close that ruminants such as cattle or deer (which have no upper incisors and bite against a horny pad in the upper jaw) cannot themselves utilize those patches. Neither cattle nor deer establish distinct latrine and grazing sites. Both dung wherever they happen to be at the time – and since they spend most of their time grazing within the pony latrines, their dung, too, accumulates in these latrine sites.

This account is of course oversimplified to a degree. Over the winter when the available grazing is at a minimum (and the numbers of domestic stock upon the Forest are also reduced so that the accumulation of fresh dung is low) the ponies do begin to graze into the previous latrine areas. By the end of winter the clear mosaic of latrine and non-latrine patches within a grassland is less apparent and the ponies feed and defaecate freely all over the lawns. However, the pattern begins to establish itself again from April/May and by June complete separation of latrine and grazing areas is re-established. Over time this dislocation in nutrient return even within a single community, leads to continued impoverishment of pony-grazing areas and continuous nutrient enrichment of latrines. Already, in grasslands ploughed and reseeded after the Second World War, consistent differences are recorded in the potassium and phosphorus content of soils, with nutrient levels being higher in the latrine areas by a factor of about 1.2 (phosphorus) to 1.7 (potassium) (Putman *et al.* 1981). Organic matter content of latrine areas is also consistently a little higher.

Differences in nutrient status and grazing regime experienced (plants growing in pony grazing areas are subjected to a closer cropping than those in latrine areas foraged only by cattle or deer) have led already to significant differences in species composition between latrine and non-latrine patches – establishing a fine-scale mosaic in species associations across the sward. Species such as ragwort (*Senecio jacobaea*) and thistles (*Cirsium vulgare* and *Cirsium arvense*) occur only in latrine areas, and other species are more or less abundant in latrine or heavily grazed patches (Putman 1986b, Ekins 1989).

Within the Forest woodlands the effects of grazing are even more apparent. While there are differences in the species composition of the ground flora of grazed areas, equivalent to those observed in open, grassland communities (with a similar shift towards species of prostrate growth form in heavily grazed areas; Putman 1986b, Putman *et al.* 1989), the most dramatic and obvious effects of the heavy grazing pressure can be seen within the shrub layer. Simply, there is a stark absence of understorey species such as hazel (*Corylus avellana*), birch (*Betula* sp.), blackthorn (*Prunus spinosa*), hawthorn (*Crataegus monogyna*), bramble (*Rubus* agg.) or rose (*Rosa canina*) – shrub-layer species which would be expected and which indeed form a dense understorey in woodlands just beyond the proverbial Forest Fence where grazing pressure is more moderate. The shrub layer in effect is represented only by the unusually abundant hollies (*Ilex aquifolium*) – in themselves a classic illustration of the separate phenomenon of competitive release: that under intense grazing pressure, while palatable species will ultimately be eliminated, the release of competition may permit an expansion in range and abundance of species which are very tolerant to defoliation, or have specific defences against attack; escaping major impact through growth form, or resisting herbivory with spines, thorns or chemical defences.

Such changes in species composition and growth form, as well as the continued imposition of grazing, have clear effects on the physical structure of these same communities. New Forest grasslands boast little vegetational material higher than a few millimetres even in the height of summer (Plate 7), for taller vegetation is immediately cropped; such grasslands clearly lack many of the possible structural layers of mature, ungrazed grasslands. On heathlands, too, the effects of grazing are very clear in this reduction of structural diversity. Heathlands in southern Hampshire support dense stands of purple moor-grass (*Molinia caerulea*) among the *Calluna* and *Erica* heaths. Outside the Forest boundary, the tall flower spikes of the *Molinia* tower above the canopy of the heather plants and provide a whole additional structural element within the vegetation. Inside the Forest, this whole stratum is missing. *Molinia* plants are just as abundant within the heathlands, but *Molinia* is an important component of pony diet and the plants are always heavily grazed – and kept well below the heather canopy. And once again the change in species composition and the continued effects of browsing are strikingly obvious in the three-dimensional structure of the Forest's open woodlands. The effects of centuries of heavy browsing pressure are so marked that New Forest woodlands virtually lack **any** ground, field or shrub layer; and indeed the whole structural 'layer' between ground level and the clear browse horizon at 1.8 m, is almost completely missing (Plate 8). Under continuous browsing pressure all the more palatable shrubby species such as hawthorn, blackthorn, hazel or willow have been eliminated and fail to regenerate. Even the resistant hollies are thoroughly browsed up to the 1.8 m browse line and bear little foliage below this level; shrubs of holly or gorse which fall entirely within the reach of the herbivores are severely stunted and 'hedged' by the continuous browsing pressure. At the ground level, brambles, ivy and other field-layer species are completely lost; the only species that gives any structure at this level is bracken (*Pteridium aquilinum*) which, although eaten by the ponies at certain times of year, is not particularly palatable!

With this obvious change in woodland architecture comes another, less apparent. Under the centuries of heavy grazing pressure, not only have many species of the field and shrub layers been eliminated: there has also been a virtual lack of regeneration of any of the canopy tree species. With pigs turned out to help the deer clear up the mast, and deer and ponies to graze upon such tree seedlings as do germinate, few trees survive beyond the seedling stage. George Peterken and Colin Tubbs (1965) noted that in consequence the Forest woodlands presented a most peculiar age structure: composed largely of trees established in the 1750s, others recruited in the mid-1850s and a third cohort established in the 1930s. This bizarre and wildly distorted age structure clearly reflects periods of reduction in browsing pressure, sufficient in most years to suppress any natural regeneration of these ancient woodlands at all, but reduced in 1750 (when new oak plantations were planted, and enclosed for protection, to provide timbers for the Navy's ships), after 1851 (following an Act of Parliament providing for the 'removal' of all deer from the Royal Forest) and during the Great Depression of the 1930s. (Political commentators might like to observe that there is now a further phase of regeneration dating from the mid-1980s!)

2.3 EFFECTS OF GRAZING ON THE FOREST FAUNA

All these various effects upon the vegetation – upon species composition, productivity, physical architecture – clearly affect the resources offered to other animals dependent on that same vegetation for food, shelter or cover from predators. The diversity and species composition of invertebrates in any community have been shown to have a strong correlation with diversity of vertical structure within the vegetation as well as spatial heterogeneity (Southwood *et al.* 1979); clear responses to a change in grazing regime have been recorded in the invertebrate fauna characteristic of particular communities, particularly among beetles and butterflies (e.g. Putman *et al.* 1989). Similar effects are recorded in other animal groups. The New Forest again offers an excellent example of the knock-on effects of the vegetational changes resulting from centuries of heavy grazing on populations of mice, voles and shrews. Comparisons of the species diversity and population sizes of these small mammals within the Forest with those recorded in equivalent vegetation types in areas outside the Forest boundary (grazed by deer but not by domestic livestock) reveal striking and consistent differences (Hill 1985, Putman 1986b, Putman *et al.* 1989).

All ungrazed woodland areas studied supported substantial populations of woodmice *(Apodemus sylvaticus)* and bank voles *(Clethrionomys glareolus)*, with lower densities recorded of yellow-necked mice *(Apodemus flavicollis)*, and both common and pygmy shrew (*Sorex araneus, S. minutus*). Rodent communities of grazed woodlands within the Forest were characterized by healthy populations of woodmice – maintaining equivalent density and dynamics to those recorded in ungrazed sites – but all other species were rare or absent.

Forest heathlands and grasslands are equally profoundly affected by grazing; the physical structure of these open vegetation types displays, as we have already noted, marked contrast with heathland or acid-grassland sites not subject to heavy grazing, and the reduced structural 'depth' provides scant cover from predation. While heathland plots beyond the Forest boundary supported large, permanent populations of woodmice and harvest mice (*Micromys minutus*), and grasslands in turn supported strong populations of woodmice and field voles (*Microtus agrestis*), small mammals were virtually completely absent from open communities within the Forest itself (on heathlands for example, the only animals caught in 1200 trap-nights were two woodmice – obviously themselves in transit).

Responses to grazing of this kind illustrate very clearly that an increase in grazing intensity in any ecological system will have implications far beyond the immediate consequences for the vegetation itself or upon direct competitors. These 'knock-on' effects have repercussions indeed throughout the entire community: the effects of grazing may be seen to have consequential effects on the abundance and behaviour of higher-order predators, neither directly linked to the dominant herbivores, nor themselves directly affected by the changes in the vegetation, but influenced by secondary changes in the abundance of prey or competitors. Within the New Forest, the reduced diversity and overall abundance of small mammals in the heavily grazed woodlands and open 'wastes' have been shown, in their turn, to have had an effect on the foraging behaviour, diet, population density and breeding success of such diverse predators as foxes, badgers, buzzards, kestrels and tawny owls (Putman 1986b).

Two independent studies of the diets of foxes (*Vulpes vulpes*) within the New Forest (Farley 1986, Senior – unpublished data quoted in Putman 1986b) reveal that while New Forest animals did consume small rodents when available, the frequency and relative proportion in the diet was lower than that recorded in other areas: few birds were taken and the foxes clearly relied heavily on invertebrate material (particularly earthworms and beetles), carrion and autumn fruit. Perhaps in response to scarcity of prey, the overall density of foxes within the Forest is also unusually low: estimated as only 2 per km^2, with an adult density of 0.75 per km^2 (Insley 1977).

Avian predators are also affected by the low rodent abundance. Colin Tubbs (1974, 1982) first noted that there seemed to be a close correlation between the breeding success of buzzards *(Buteo buteo)* in the New Forest and population density of grazing cattle and ponies. Tubbs noted that in the New Forest, as in other parts of England where rabbits are not readily available, buzzards appear to rely very heavily upon rodent prey, and breeding success is directly related to the abundance of such rodent prey. Although he had no direct data on the changing abundance of small mammals within the Forest over the years, he none the less showed a clear correlation between the number of buzzard pairs attempting to breed in any year and the numbers of domestic stock grazed on the Forest over the preceding 3 years (Tubbs and Tubbs 1985; Figure 2.2).

Similar effects have been demonstrated by Graham Hirons, who examined diet and breeding success of other raptors within the Forest whose diets would normally be expected to contain high numbers of rodents. Tawny owls *(Strix aluco)* – very much woodland predators – continued to maintain a high proportion of rodent prey within the diet, although this was almost exclusively woodmice. However, this contributed only 42% of all prey taken (as against 60–70% recorded from studies elsewhere) and New Forest owls compensated with increased reliance on invertebrate prey, particularly dung beetles. Hirons noted a significant reduction in density of owls overall and a reduction also in the proportion of pairs breeding in any year (25% in the Forest as against 65% of pairs in areas outside; Hirons 1984).

None of these effects of grazing is of course unique to the New Forest: equivalent examples of the effects of grazing upon the structure and species composition of vegetation are legion, and extensively reviewed elsewhere (e.g. Gessaman and MacMahon 1984, Putman 1986b). 'Knock-on' effects of such changes upon other herbivores or their predators are also increasingly commonly reported. (Research undertaken by the

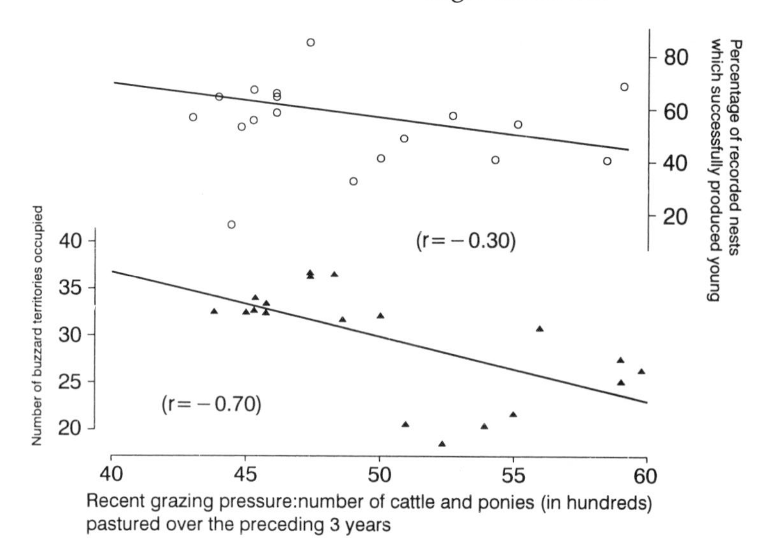

Figure 2.2 Relationship between grazing pressure on the New Forest and breeding success of buzzards (based on data from Tubbs and Tubbs 1985). Note that while success of nests, once a breeding attempt has started, shows no significant effect of grazing pressure, the actual number of buzzard pairs attempting to nest declines as grazing pressure increases.

Forestry Commission for example (Petty and Avery 1990) has shown that immediately after the cessation of grazing in upland areas about to be afforested, populations of field voles show an initial surge in population density. This increased food supply is exploited by predators such as short-eared owls *(Asio flammeus)*, long-eared owls *(Asio otus)*, kestrels *(Falco tinnunculus)* and barn owls *(Tyto alba)*, whose breeding success and population sizes also increase until canopy closure within the new plantations.) But what is perhaps unusual is that in the New Forest all these various changes are clearly documented within a single system.

And it is this detail of understanding, this continuity of concentrated work within a single area that also makes this particular system an ideal one in which to explore the interactions between the forest herbivores themselves.

2.4 THE FOREST'S LARGE HERBIVORES AND THEIR MANAGEMENT

As noted, the New Forest area currently supports populations of four main species of wild or feral deer (fallow, roe, red and sika: five species if we now include Reeves' muntjac as a recent colonist), besides substantial populations of free-ranging cattle and ponies. (As muntjac deer are perhaps a minor character among the cast of the wild or feral species, small numbers of sheep and donkeys are also pastured under Common Rights in small areas of the Forest – but these too may be regarded as having only a 'walk-on' part in our drama.)

Fallow deer (*Dama dama*) have long been the dominant deer species within the Forest; indeed William I's declaration of the area as a Royal Forest was chiefly to conserve hunting interests for this species. It is difficult to assess what numbers may have occurred in the Forest at that time. The earliest complete 'census' is that of 1670, when the Knights Regarder charged with administration of the Forest returned an estimate of 7593 fallow deer and 357 red deer within the Forest boundaries. (Painfully aware myself of the difficulties of assessing population sizes of any deer species in a huge area of difficult terrain, the lengthy debates in the literature about the accuracy and application of alternative methods – and the huge confidence limits one places around even approximate estimates, I cannot help but marvel at the delightful precision of these figures: 7593 – and take them with a pinch of salt!) A government report of 1789 gave a more global estimate of the average number of fallow deer present as 5900 and numbers seem to have remained at roughly this same level until the 1850s. In 1851, the New Forest Deer Removal Act, in relinquishing the Crown's rights to an exclusive hunting reserve, provided for the 'removal' of all deer from the Forest within 3 years of the enactment. Total extermination of such a large population of animals, scattered over so large an area, was of course as impracticable as a precise count, but numbers were certainly dramatically reduced and population estimates in 1900 gave a figure of 200 head (Lascelles 1915). Since that time the population has gradually expanded and is now maintained by culling at a level which has been estimated at about 2000 animals (Strange 1976, Putman and Sharma 1987).

Red deer (*Cervus elaphus*) were also established in the New Forest at the time of William the Conqueror, but numbers were always substantially lower than those of fallow, and the population throughout seems to have been barely self-sustaining. Populations have indeed been continuously 'subsidized' by introductions – not merely in an attempt to improve the perceived 'quality' but also simply to bolster numbers. Both James I and Charles II (reigned 1603–25 and 1660–85, respectively) introduced fresh blood from France, Charles II importing no fewer than 375 red deer which were released near Brockenhurst in the south of the Forest. Further introductions continued throughout the nineteenth century and even into the early twentieth century.

Census records are patchy; during Henry VII's reign (1485–1509) there were several records of red deer being killed within the Forest; the Regarders' survey of 1679 (above) estimated numbers at 357: 103 male and 254 female deer. By the late eighteenth century however, the Forest's red deer population was almost certainly extinct; certainly returns of 1828–30 of deer in Royal Forests omit any mention of red deer within the New Forest. References to sightings in the nineteenth century probably relate to escapes from nearby deer parks, and in the past 200 years numbers have probably never exceeded 80–100 animals.

Roe (*Capreolus capreolus*) may also be presumed to have been native in the New Forest area, but by medieval law, red deer were beasts of the Forest (reserved for Royalty) while fallow and roe, the lesser beasts of the 'chase' were generally less jealously protected – and during the Middle Ages roe deer became virtually extinct throughout England and much of Scotland. As with the red deer, modern populations of roe in southern England have resulted from reintroductions of animals into several areas during the nineteenth and twentieth centuries (Prior 1973). Roe recolonized the New Forest from 1870 onwards, spreading across from Dorset (Jackson 1980). Census figures suggested a population of perhaps 400–500 animals in the early 1970s, but since that time numbers have declined substantially (see Chapter 6). Numbers are now extremely low (estimated in 1990 at *c*. 300) and somewhat patchily distributed within the Forest.

Sika deer (*Cervus nippon*) are an even more recent and completely exotic introduction to the New Forest community. Sika were first introduced to Great Britain in the 1860s (reviewed by Ratcliffe 1987a) and to the New Forest in the early 1900s. Current populations are descended from animals escaping from a collection in the nearby Manor of Beaulieu, and for many years were restricted to a small area in the south of the Forest, an area seemingly bounded to the north by a major railway line, by the sea to the south, and to the east and west by the waters of the Beaulieu and Lymington Rivers. But these boundaries were hardly impassable, certainly those to the north and west, and although populations were contained until perhaps the late 1970s or early 1980s, as numbers within the core of this curiously restricted range increased so, by the late 1980s, increasing reports were made of sika spreading beyond this initial localized area. By the late 1980s numbers were assessed as in excess of 200–300 animals and while still restricted to the southern part of the Forest, sika were recorded over a far more extensive range. Numbers were reduced by heavy culling in the late 1980s and are now held at approximately 100.

As is apparent from all of this, numbers of all deer species within the Forest are artificially controlled. Responsibility for management of the Forest's deer population passed to the state Forestry Commission in 1924. Since 1960 regular annual census of populations has been attempted; on the basis of this routine monitoring of population trend, cull quotas are established in an attempt to maintain numbers of each species at predetermined levels judged to be compatible with maintaining healthy populations of the animals, yet at levels consistent with other demands upon the area – for timber production, recreation and conservation.

While multiobject use of our forests and other 'wild' spaces is very much the current vogue of management, and certainly offers a more balanced and sensitive approach than the simple coniferous blanket of pure commercial forestry, the management of the New Forest must long have been a nightmare. Ever since administration passed to the Forestry Commission in 1924, it has had to try and balance the interests of commercial forestry, conservation and amenity in an area now designated not only a Site of Special Scientific Interest, but of National Park status. And one of the added complications in this thankless exercise of trying to balance all interests is the fact that alongside all its other uses, the New Forest has for centuries supported a traditional system of Common agriculture – with rights of pasturage of domestic animals, centuries old, still jealously guarded and widely exercised today.

Before its afforestation in 1069 the open wastes of the Forest would have been freely grazed by local cottars and farmers. Pigs would have been turned out routinely in autumn to feed on the acorns and beech mast. Timber, turves or peat would have been cut from the Forest for fuel, and bracken cut for bedding or litter. After the imposition of Forest Law, these practices were allowed to continue, but under strict control and regulation – eventually, within the New Forest, being defined in the New Forest Acts of 1689. Common Rights could be claimed by an cottager or landowner holding more than an acre of land within the Forest boundaries and actually embrace a number of different entitlements besides those of pasturage and pannage. Thus these rights included Right of Common of pasture (for cattle, ponies and donkeys); Right of Common of mast (the right to continue to turn out pigs during the mast season, or 'pannage'); Right of Common of turbary (the right to take turf fuel); Right of Common of estovers (the right to take fuel wood); and Right of Common of marl (the right to take marl from recognized pits within the Open Forest). To these may be added the right to cut 'fern' (bracken) for bedding and litter, and sheep rights claimed by the adjoining manors of Beaulieu and Cadland.

Rights of pannage and Common pasturage were undoubtedly the most significant and important rights which could be claimed by Forest Commoners in terms of maintaining a livelihood; it is clear that many of the holdings were of such a size that they would not have been viable without these rights upon the Forest. While most claims to pasturage recorded in the initial Register of Claims in 1635 conceded the old rules under Forest Law of levancy and couchancy (limitation of the number of stock turned out in the spring and summer to that number which could be maintained on the holding over winter) this was clearly not rigidly adhered to, nor is it today.

Administration of the Forest grazing and oversight of Common Rights is carried out today by a specially appointed Court of Verderers, whose statutory responsibilities are to make by-laws for the control of stock upon the Forest, to safeguard the grazings and the health of the Common stock (through their appointed Agisters), and to maintain the Register of Common Rights. To defray the necessary costs of administration – as well as the employment of four full-time Agisters responsible for the day-to-day welfare of the Common stock turned out upon the Forest – the Verderers levy a 'marking fee' for each animal licensed to be turned out: and in effect the limit to the number of stock depastured by any Commoner is set by the number of marking fees he is prepared to pay rather than by any adherence to ancient laws of levancy and couchancy. Thus there is nowadays no formal limit to the total number of stock licensed to graze upon the Open Forest at any one time – and that limit is in effect determined by the Verderers on the basis of the numbers they think the Forest can effectively support.

While Registers of Claims to Common Rights on the New Forest have been compiled at regular intervals since the seventeenth century, such records cannot show the extent to which these rights were exercised. Thus it is difficult to offer precise figures for the number of Common stock pastured on the Open Forest at any given juncture. A census of stock carried out in 1895 gave a figure of 2903 ponies, 2220 cattle and 438 sheep, but since that time one can only document trends in the overall numbers of marking fees paid, which merely establishes the maximum number of stock licensed on the Forest, rather then the actual number turned out at any one time.

By 1910, the numbers of ponies for which marking fees had been paid had fallen to 1500; cattle numbers remained at between 2000 and 2500. In 1920 numbers rose again to a peak of 4450 stock (mostly cattle) before the Depression led to a slump in Commoning practice – and by 1949, at the end of the Second World War, there were only 1757 domestic animals licensed to graze upon the Forest. Numbers pastured over more recent years have also fluctuated considerably, depending largely on economic factors. Numbers over recent years are summarized in Figure 5.1, from which it may be seen that after a peak in total numbers through the late 1970s and early 1980s, numbers have declined again more recently, to around 3000 ponies and 1600 cattle by 1988.

2.5 CURRENT POPULATIONS

In practice, numbers of animals of any species on the Forest are very hard to estimate: licensing fees do not represent for Common stock what may actually be turned out at any one time. Even for the deer our estimates are imprecise. The annual census undertaken by the Forestry Commission is designed to monitor trends in population rather than assess actual numbers. For management purposes, the whole Forest is divided into 12 'beats' (ranging in size from 1470 to 2710 ha). Each beat is the responsibility of one keeper, who makes a return to the central office each year of the number of deer of each species held within his beat on a fixed date in April. In practice this figure is reached as a compromise between the keeper's accumulated knowledge of the number of deer in the area over the previous few weeks and an actual 'single-day' census on the appointed date. Such visual counts consistently underestimate population size in concealing habitats, so that the population estimates returned cannot be assumed to be accurate.

Censused figures for fallow deer within the Forest over the period 1970–75 (Table 2.1) suggest a total population of 914 fallow deer within the New Forest over that period (mean over the 6 years). Calculating the total population required to sustain imposed mortality (culling, road traffic accidents and known poaching losses) given actual recruitment rates (fawns/doe) Strange (1976) considers the true population at that time was probably nearer 1800. Mann's (1983) figures for sika deer populations within their core range between 1979 and 1982 suggest numbers of between 175 and 200; once again such an estimate is double the figure given by the Forestry Commission census for the same period of 74–95. The population size for red deer is similarly consistently underestimated (Payne 1987), although it is probable that the census figure for roe (which in the New Forest are more characteristic of more open ground) more closely approximates to the true total.

We may suggest therefore that the total New Forest area supported populations of perhaps 2000 fallow deer, 200 sika deer, perhaps 120–150 red deer and between 250 and 300 roe deer over the period considered in the analyses presented in these pages; over that same period the 20 000 hectares of unenclosed land supported in addition between 5000 and 6000 cattle and ponies.

2.6 REPRISE

The New Forest is truly a remarkable area: remarkable as one of the largest areas of semi-natural vegetation in Britain, remarkable for the variety of its vegetation types – some of which, like the valley mires, are recognized amongst the finest in Europe. It is remarkable too for the richness of its history, for the persistence to the present day of widely exercised rights of Common pasturage, and as a dramatic example of the ultimate impact of a continuous history of heavy grazing upon the structure and composition of different communities. Not surprisingly, therefore, it has attracted a great deal of interest and research effort (see again

Table 2.1 Changes in censused numbers of New Forest deer populations 1960–92. Figures presented are based on the annual census carried out by the Forestry Commission and offer an index of population changes.

	Red	*Sika*	*Fallow*	*Roe*
1960	–	35	788	211
1961	–	44	857	302
1962	7	34	931	328
1963	23	42	933	383
1964	20	64	994	434
1965	23	84	1147	465
1966	24	97	1036	490
1967	29	67	968	527
1968	33	38	855	539
1969	18	38	893	572
1970	23	37	1022	620
1971	26	71	866	444
1972	28	73	1017	428
1973	29	75	904	435
1974	37	78	909	356
1975	49	88	834	375
1976	43	94	973	344
1977	49	80	908	363
1978	79	79	927	332
1979	34	80	950	354
1980	58	74	1016	334
1981	73	95	996	309
1982	64	76	1049	336
1983	72	97	1020	281
1984	57	115	1033	264
1985	66	99	1006	260
1986	70	103	1086	265
1987	77	103	1227	249
1988	77	79	1152	265
1989	94	90	1280	255
1990	145	78	1180	295
1991	97	101	1204	351
1992	93	96	1290	369

Tubbs 1968, 1986, Putman 1986b, for general reviews). Amongst this, considerable attention has also been directed towards the behaviour and ecology of the Forest's larger herbivores. Although Jackson's pioneering work on the fallow deer of the Forest (Jackson 1974, 1977, 1980) was undertaken in the earlier part of the decade (1970–73) and Tyler's detailed analysis of the social organization of the Forest ponies (Tyler

1972) was also undertaken some years previously, really concentrated research effort on the ecology of the Forest's large herbivores began towards the end of the 1970s.

At this time, the numbers of domestic livestock grazing on the Open Forest was increasing steadily (Figure 5.1); the Nature Conservancy Council, concerned about the possible impact on the Forest vegetation of excessive grazing, commissioned a study of the foraging ecology and behaviour of domestic cattle and ponies and their impact on the vegetation of the New Forest (Putman *et al.* 1981, 1984, 1987, 1989, Pratt *et al.* 1986). Further work followed on the ecology and behaviour of the Forest ponies and their ability to maintain condition as stocking density approached capacity (Gill 1988, 1991, Burton 1992); Jackson's early work on the ecology of fallow deer was extended (Thirgood 1991, 1995a,b, Putman *et al.* 1993, Parfitt, unpublished) and detailed studies initiated of sika deer (Mann and Putman 1989a,b, Putman and Mann 1990), red deer (Payne 1987) and roe (Sharma 1994).

The wealth of data accumulating on each species in isolation offered tremendous potential for looking at both direct and indirect interactions between the different ungulate species themselves. And while our early work within the Forest was not deliberately directed towards such analysis of interaction, a number of more recent studies have sought explicitly to explore those interactions between the different species of large herbivore: explicitly looking at habitat use and diet of pairs or groups of species in direct sympatry (Boxall 1990, Sharma 1994).

These studies offer tremendous potential in our present endeavour: the detailed analysis offered of the behaviour and resource use of each species in this complex assemblage, together with explicit analyses of interactions between those species, provide us with a remarkably well-researched 'model system' through which to explore our more general question of the dynamics of coexistence in multispecies ungulate assemblages.

3

Ecology and behaviour of the Forest's fallow deer

3.1 INTRODUCTION

Although widespread throughout Europe some hundred thousand years ago, fallow deer probably became extinct during the last glaciation except for a few small refuges in southern Europe. Their secondary radiation from these relict populations to the rest of mainland Europe and to Britain was at least assisted by man (Chapman and Chapman 1980, Chapman and Putman 1991). Indeed, of all the deer of the world, fallow are perhaps the species whose current distribution has been most influenced by man; further introductions have taken this species well beyond Europe and it is now established in the wild in both North and South America, Africa and Australasia, as well as in its native Eurasia.

It is thought that fallow were brought to the British Isles by the Romans or Normans; the species is now widely established in the feral state within the UK, with numbers estimated by Harris *et al.* (1995) at around 100 000. Fallow have long been the dominant deer species within the New Forest – and it is clear that the original designation of the area as a Royal Forest was chiefly to conserve hunting interests for this species. In this chapter we will review the behaviour and ecology of fallow deer within the New Forest: concentrating specifically on distribution, movements, habitat use and feeding behaviour.

There should be little need to go into too much detail at this point on the general biology of fallow deer. An excellent review of the species was presented by Chapman and Chapman (1975) – and indeed fallow deer form the focus of another book in this current series (Langbein and Thirgood, in press). However, for readers unfamiliar with the general natural history of the species it may be appropriate to offer a brief introduction here. In the wild state, fallow deer are characteristic of mature woodlands; although the deer will colonize coniferous plantations, provided these contain some open areas, they prefer deciduous woodland with an established understorey (Chapman and Putman 1991). These woodlands need not necessarily be very large, because they are used primarily for shelter. Although the deer may feed within the cover of the woodland, particularly at the time of the autumn mast crop, or over winter when the diet includes a greater proportion of woody browse, they are not, in practice, dependent on the woodland's food supplies. Fallow are preferential grazers (Jackson 1977, Caldwell *et al.* 1983, Putman *et al.* 1993) and, while in larger woodlands they may feed on grassy rides or on ground vegetation between the trees, they will as frequently leave the trees when feeding to invade agricultural land or other open land beyond the woodland edge (Thirgood 1995a).

Traditionally regarded as a herding species, fallow have a rather complex social system and social organization is closely linked to the annual cycle. In larger woodlands, males and females remain separate for much of the year, with adult males observed in bachelor groups and females and young (including males to 18 months of age) forming separate herds, often in distinct geographical areas. Males come into female areas early in autumn to breed; mature bucks compete to establish display grounds in traditional areas and then call to attract females. From this time on, adult groups of mixed sex may be observed through to early winter. Rutting groups then break up and the animals drift away to re-establish single-sex herds.

However, such a picture is greatly oversimplified. In common with that of many ungulates, fallow deer social organization seems to be extremely flexible and strongly influenced by environment (see Langbein and Thirgood in press). Both group size and degree of sexual segregation are profoundly affected by habitat (Putman 1988, 1993, Thirgood 1995b) and in more open environments, such as agricultural areas, adult groups of mixed sex may be observed throughout the year (Thirgood 1995b). Reproductive strategy is equally variable, with males in some areas establishing the traditional, discrete, 'rutting' stand or breeding territory, others defending only temporary stands and thereafter consorting with the harem they have attracted, others again merely foraging throughout their range for oestrus females (Langbein and Thirgood 1989, Thirgood 1991, Putman 1993). Fallow males may also, under special circumstances, establish breeding 'leks' analogous to those of other lek-breeding birds and mammals (Clutton-Brock *et al.* 1993); such lek-breeding populations have attracted a great deal of recent interest (Schaal 1986, Schaal and Bradbury 1987, Clutton-Brock *et al.* 1988, 1989, Apollonio 1989, Apollonio *et al.* 1989, 1990, 1992). In effect, however, the form of mating system adopted is determined primarily by population density and environmental character (Langbein and Thirgood 1989, in press) in much the same way as the observed variation in grouping patterns or sexual segregation (Putman 1993, Thirgood 1995b).

3.2 SOCIAL ORGANIZATION

Within the mixed vegetational mosaic of the New Forest, sexual segregation between males and females outside the breeding season is extremely marked. Bucks and does characteristically occur in distinct and separate groups, and although in certain areas the actual geographical range of these 'bachelor' and 'small-deer' herds (females, fawns and followers) may overlap to an extent, over much of the Forest even these social groupings are geographically segregated so that one may recognize distinct 'buck' and 'doe' areas. Except during the autumn rut (October/November) over 95% of all groups of fallow deer encountered within the Forest are either exclusively male (containing adult males, younger bucks and yearlings) or exclusively 'small deer' (females and followers; such groups may include yearling males which have not yet left the natal group); groups containing adults of both sexes account for less than 5% of all groups observed and appear to be largely a result of young males (of between 2 and 4 years old) joining female feeding aggregations (Thirgood 1990, 1995b).

Thirgood notes, however, that analysis of sexual segregation at the level of the group, while accurately reflecting the experience of the human observer, may not so accurately reflect the experience of the individual animal. He notes for example (Thirgood 1990) that in a population where females greatly outnumber males, if only 10 males are observed out of a total sample of 1000 deer, and all of those males are actually in mixed sex groups, which in total comprise 100 deer, the estimate of population segregation of 90% does not reflect the experience of the typical male. He argues that group-based methods of assessing segregation will generally suggest higher levels of segregation than those based on individual experience and offers an additional analysis of sexual segregation, by assessing (as Schaal 1982) the proportion of individual deer occurring in single-sex or mixed-sex groups. Such analysis (Thirgood 1990, 1995b) still suggests that (outside the period of the autumn rut) more than 90% of females will be found in single-sexed herds, but that for males a higher proportion of adults (19.5%) form part of mixed-sex, rather than purely male, groups (Thirgood 1995b).

Fallow are commonly regarded as one of the more social of the deer species and certainly may frequently be encountered in large aggregations. However, such sociality is somewhat illusory (Putman 1981, 1993) and derives at least in part from animals that occupy overlapping home ranges coming together in favoured areas to feed. More typically, fallow in the New Forest occur in far smaller social groupings of between one and five individuals (Jackson 1974, Parfitt in Putman 1986b, Thirgood 1995b). The distribution of size classes in mixed-sex groups is more uniform, with the larger size classes occurring more frequently (Thirgood 1995b); such an observation might, however, be expected since a large number of such mixed-sex groups will actually have formed as temporary aggregations of animals of more than one social group on favoured feeding grounds.

However, as in our earlier analyses of sexual segregation, description of the frequency distribution of observed groups of particular size, while describing accurately the distribution of the population into units of given size, may not so accurately reflect the experience of the individual animals within that population. A group of 100 individuals is a single group, but actually represents the social experience of 100 separate individual animals. Jarman (1974, 1982) has emphasized the experience of the individual as the biologically more meaningful representation of grouping for any social species and has introduced the weighted measure of 'typical group size' (as the group size experienced by the typical individual) rather than simple mean group size (the arithmetic mean of observed group sizes) as an alternative method for assessing grouping tendency. Such perspective is of course itself heavily skewed by even infrequent encounters of groups or aggregations of large size, but this alternative view of group-size distribution suggests for New Forest fallow deer a more even distribution of grouping patterns overall than that suggested above, with individual deer observed equally frequently in larger group sizes (Table 3.1; Thirgood 1995b).

Table 3.1 Group sizes of New Forest fallow deer: (a) Frequency of observation of groups of different sizes (presented as percentages of all groups seen); (b) percentage of individual deer recorded in groups of different size. (Calculated from data of Thirgood 1995b.) N = number of observations.

(a) Group sizes observed:

	1–3	4–6	7–9	10+	N
Males	72.0	10.3	7.3	10.3	465
Females	65.3	20.8	6.8	7.2	2496
Mixed sex	28.9	23.3	17.1	30.7	322

(b) Group sizes experienced:

	1–3	4–6	7–9	10+	N
Males	30.2	13.4	16.2	40.2	1759
Females	27.4	24.0	12.5	36.1	10470
Mixed sex	7.6	11.6	13.4	67.4	3210

Clear seasonal variation in both size and composition of groups is reported by all these authors and group size may also be seen to vary with habitat occupied (Putman 1981, 1993, Thirgood 1995b). Figures 3.1 and 3.2 show the frequency of occurrence of groups of particular size amongst male deer and groups of 'small deer' (females and followers) in John Jackson's study areas, and seasonal changes in the frequency with which a particular group size might be encountered (Jackson 1974). Almost identical patterns were reported by Parfitt (in Putman 1986b) and Thirgood (1990, 1995b). From November to January, almost all males were encountered singly or in pairs; in February and March, groups of three to five were almost equally commonly observed, but from then on these larger groupings became progressively less frequent through the summer and autumn until by November, almost all males were again encountered only in ones or twos (Jackson 1974, see also Thirgood 1990). Females show more variation in group size throughout and are almost equally likely to be encountered in groups of one or two, or three to five in all seasons (Figure 3.2). Larger social groups are, however, more commonly observed in April and May; during late May the groups break up as females become progressively more solitary in preparation for the birth of their fawns in mid June. By August, numbers in herds increase again as does and fawns join with other family groups, rising again in September and October as females collect at the rutting stands in autumn.

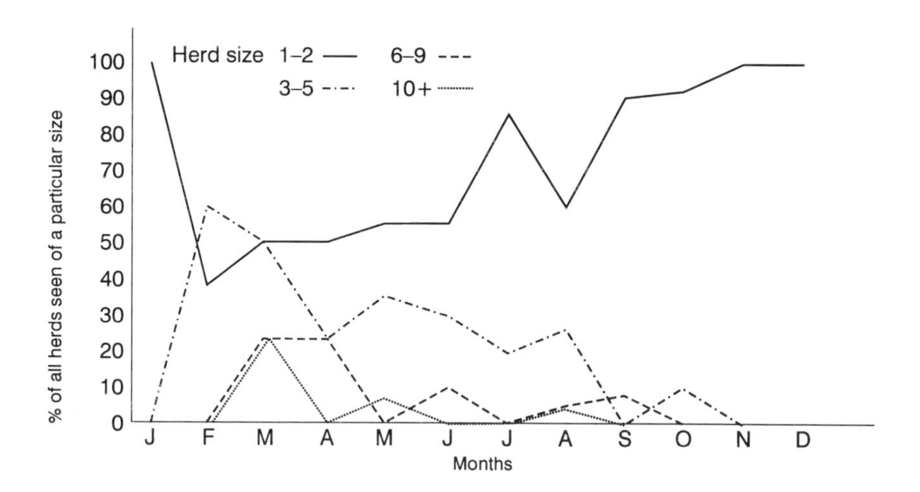

Figure 3.1 Seasonal fluctuations in the group sizes of New Forest fallow bucks (from Jackson 1974, with permission).

Figure 3.2 Seasonal fluctuations in the recorded size of groups of fallow deer females and followers within the New Forest (from Jackson 1974, with permission).

Parfitt (in Putman 1986b) detected little variation in group size with habitat; in his analyses no consistent differences in group sizes of small-deer groups could be detected between habitats (sample size of male groups was too small to permit any formal comparison) and the dominating influence on changing group size remained the seasonal effects of the annual cycle. However, Parfitt's observations were restricted in the main to woodland Inclosures and perhaps did not offer a sufficient variation in structure between different habitats to detect a response in social behaviour. Thirgood (1995b) has subsequently demonstrated significant differences in group sizes formed in different habitats. Once again, sample size – and the small overall range in group size recorded – did not permit detailed analysis for males, but sizes of small-deer groups, whether assessed from the perspective of size of group encountered by the observer or the group size of the typical individual animals, were approximately 2–4 times larger in more open habitats than within woodland (Table 3.2). There appeared to be little difference in group sizes recorded in different woodland types (open- or closed-canopy woodlands, coniferous or deciduous) but group sizes on favoured grasslands were consistently higher than those on other open-community types (e.g. heathlands; Thirgood 1990).

All these data suggest that both season and habitat may have an effect on grouping patterns of male and female fallow deer; effects of habitat are further independent of any effect of season. One might consider that if group size is affected by habitat and patterns of habitat use change

Table 3.2 Effect of habitat on group size in fallow deer. Data (from Thirgood 1990) are presented from one representative site within the Forest – Denny Lodge Inclosure, by way of illustration of changes in grouping in different habitats. Group sizes in each habitat are presented (± one standard error) as both: (a) mean group size (arithmetic mean of groups seen by observer) and (b) typical group size (group size of the average individual, after Jarman 1982).

	Closed-canopy woodland	*Open-canopy woodland*	*Mixed woodland*	*Rides/ Glades*	*Grassland*	*Heathland*
Group type						
Males						
(a)	1.0	2.1 (0.32)	2.0 (0.71)	2.5 (0.83)	2.4 (0.30)	2.8 (0.64)
(b)	1.0	2.8 (0.16)	2.5 (0.50)	3.6 (0.52)	2.7 (0.15)	3.7 (0.39)
Females						
(a)	2.4 (0.32)	3.5 (0.16)	2.8 (0.12)	2.3 (0.13)	5.3 (0.44)	7.2 (0.98)
(b)	3.4 (0.26)	5.7 (0.12)	4.2 (0.15)	3.7 (0.13)	8.7 (0.13)	10.0 (0.34)
Mixed sex						
(a)	4.0	6.6 (0.89)	2.0	6.7 (1.7)	10.1 (2.2)	5.5 (0.61)
(b)	4.0	8.9 (0.32)	2.0	9.7 (0.74)	14.5 (0.54)	5.9 (0.26)

seasonally (p. 39) then observed seasonal patterns in group size might merely reflect these seasonal changes in habitat use; conversely, apparent changes in group size in relation to habitat could result from the combined effects of seasonal changes in grouping patterns and habitat use. In practice, however, differences in group size recorded in different habitats are consistent in all seasons (Thirgood 1990, 1995b). This habitat-related variation in group size has also been shown by Heideman (1973), Schaal (1982) and Waterfield (1986), and reflects a general trend amongst ungulates to aggregate in open environments (e.g. Walther 1972, Hirth 1977, Jarman and Jarman 1979, LaGory 1986, Putman 1988); in a comparison of social structure of fallow deer in the New Forest with that of fallow deer of agricultural landscapes, Thirgood (1995b) found that although there was considerable overlap in grouping patterns in the two environmental types, group sizes were in general higher, and the degree of sexual segregation generally lower, amongst populations of agricultural sites by comparison to those observed within the New Forest.

New Forest fallow deer also featured prominently in early analyses of the variability of reproductive behaviour in this species (Langbein and Thirgood 1989). Although conventional wisdom (e.g. Chapman and Chapman 1975) describes for fallow deer a relatively invariable mating system whereby adult males establish in early autumn a mating territory ('rutting stand') within areas frequented by female herds, a territory defended vigorously against other adult males, and to which oestrus females are called by repeated 'groaning' of the successful male, more recently it has been recognized that this species displays a tremendous range of different mating strategies in different circumstances (Langbein and Thirgood 1989, Putman 1993). In practice, males in various populations may establish exclusive rutting stands as just described, or may share the stand with another younger male. Stands may be widely dispersed from other mating territories, or may be contiguous in clusters of adjacent stands; in the extreme, defended territories become vanishingly small and groups of males cluster together on a communal display ground or lek. Nor indeed, are all mating strategies observed territorial. In some populations males establish harems of females; a temporary stand may be established at the beginning of the rut, but once the buck has attracted a group of females, the territory is relinquished and ownership and defence is transferred instead to the harem. In other populations, where sexual segregation throughout the year is not pronounced, bucks compete for neither territories nor harems: males and females mix in large mixed-sex herds containing adults of both sexes; in such multi-male groupings, bucks seem to compete for simple dominance within the herd, with dominance rank conferring right of access to oestrus does. Finally, other males again seem to avoid competing altogether, but become wanderers, travelling widely through the females' ranges, mating opportunistically with oestrus does as they come across them.

Which of this diversity of strategies is observed in any population seems to depend on a combination of three factors: buck density, density (or absolute number) of breeding does and the amount of tree cover within the population's home range. Comparing populations where males showed territorial strategies against non-territorial populations, Langbein and Thirgood (1989) found that populations where males adopted non-territorial rutting behaviours were characterized by a rather low density of both does and mature bucks (animals of 4 years old or older; Table 3.3); as density increased, so there was a greater likelihood of males opting for territorial mating strategies. Among territorial strategies, a switch between individual territoriality and multi-male displays appeared also to respond primarily to the density of mature males in the population: with multi-male systems developed among populations of particularly high density and, especially, with high absolute numbers of mature bucks competing for matings (Langbein and Thirgood 1989, 1997).

Commonly the majority of males within a population (exposed to essentially the same environmental conditions) adopt the same fundamental mating strategy. In the New Forest, however, individual bucks adopt a diversity of different strategies (Thirgood 1991). The majority of males may be observed holding individual, defended, mating territories

Table 3.3 Environmental parameters characteristic of different mating strategies in fallow deer. Average values of buck density, doe density and tree cover within the home range, for populations of fallow deer characteristically showing female-oriented, territorial or multiple-male mating strategies. Data presented are derived from Langbein's studies on park fallow deer (Langbein 1991). Figures shown are mean values over a number of different parks in each case, with standard deviation in brackets.

	Mating strategy adopted by the majority of bucks		
	Harems/ Dominance within mixed groups	*Single owner rutting stands/ temporary stands*	*Leks/ Multiple stands*
Density of mature bucks (>3 yrs) per hectare	0.05 (0.01)	0.20 (0.10)	0.42 (0.04)
Average number of bucks (>3 yrs) in population	23.0 (1.4)	13.3 (9.2)	60.6 (33.5)
Density of does (>1 yr) per hectare	0.28 (0.06)	1.47 (0.36)	2.17 (0.51)
Number of does >1 yr	142.8 (27.6)	146.3 (144.0)	302.0 (155.0)
Proportion of home range with tree cover [%]	8.3 (6.9)	33.4 (12.2)	7.8 (13.2)

Note the clear differences in average densities of both bucks and does between those populations showing female-based and those showing territorial strategies; note also the differences apparent between populations adopting single-male and multiple male territories. Single-owner rutting stands are also characteristic of populations whose home range provides a high degree of tree cover, whereas multiple-male, or female-based strategies are characteristic of more open conditions.

(the 'traditional' stands), but while these are commonly widely separated from each other, in a number of instances such stands are found contiguous in space as local clusters. Males in at least two separate areas of the Forest have been observed to form leks in at least some years; finally, a number of individuals are observed adopting a completely non-territorial strategy in following groups of oestrus does around their range. Individual males are not necessarily 'specialists' in one strategy or another, but may switch between strategies between years or even within one rutting season. In observations based around one of the Forest's established lekking grounds in 1989, Thirgood (1991) notes that of 18 individually recognized males observed defending lek territories or individual stands over the 30 days of the observed rutting period, 10 were also observed interacting with females in a non-territorial strategy; males also regularly switched between lekking and single-territory defence (Table 3.4). In all, Thirgood observed the rut in 3 years (1987, 1988, 1989); in all years a large proportion of males observed were non-specialists (65% in 1987, 45% in 1988 and 42% in 1989) (Thirgood 1991).

Table 3.4 Flexibility in mating strategy of individual fallow bucks. Numbers of male fallow deer adopting specialist and non-specialist mating strategies in each of 3 years' observation at a lek-site in the New Forest (from Thirgood 1991).

	1987	*1988*	*1989*
Strategy:			
Lek only	5	3	2
Single territory (rutting stand) only	0	2	3
Non-territorial only (following)	3	6	13
Lek + single territory	4	1	3
Lek + non-territorial	4	3	7
Single territory + non-territorial	2	2	2
Lek, single-territory + non-territorial	5	3	1
Total	23	20	31

3.3 PATTERNS OF HABITAT USE

In support of our current analysis of the mechanisms permitting coexistence of so many related ungulate species within the Forest community, we should in this review turn our attention particularly to patterns of resource use. The pattern of use by fallow deer of the various habitats offered by the New Forest has been described by Jackson (1974) and Parfitt (summarized by Putman 1986b). Both concentrated their attention on populations based within fenced areas of woodland (the Forest Inclosures, from which domestic stock are theoretically excluded), rather than on the Open Forest. The deer thus had available to them a rather restricted range of habitats. Of those communities described above (Chapter 2), Parfitt and Jackson considered that the deer had ready access only to woodland habitats (deciduous woodland, mature coniferous stands, plantations and thicket areas, rides and glades). Both authors described a very similar pattern of use of these available habitats (e.g. Table 3.5).

Deciduous woodland was actively selected in early spring (February–April) and autumn (August–October). Woodland use remained high throughout the winter in good mast years, as the animals remained to feed on the abundant beech mast and acorns. Where mast was less abundant, use of the woodland blocks themselves declined over the winter and the deer made increasing use of more open habitats, grazing along rides and in clearings; rides and clearings within the woodland were

Table 3.5 Habitat use of fallow deer of Forest Inclosures (from observations of Parfitt, unpublished; in Putman 1986b, *revised*). Figures presented show the percentage of observations in any one month recorded in each available habitat.

	Jan.	*Feb.*	*Mar.*	*Apr.*	*May*	*June*	*July*	*Aug./Sept.*	*Oct.*	*Nov.*	*Dec.*
Deciduous woodland	13	34	59	43	16	20	9	39	41	11	26
Coniferous woodland	35	28	8	32	38	35	24	25	22	20	23
Rides and glades	31	30	19	21	30	30	59	26	28	42	38
Acid grassland/ wet heath	10	4	2	2	10	10	3	2	2	12	13
Grassland	0	0	0	0	0	0	0	0	0	0	0
Dry heath	0	0	0	0	0	0	0	0	0	0	0
Bog	0	0	0	0	0	0	0	0	0	0	0
Others	11	4	12	2	6	5	5	8	7	15	0

again used heavily in midsummer (June/July). Changes in habitat use recorded clearly reflect changes in the animals' need for shelter, together with changes in the degree of cover offered by different habitats in different seasons and the availability of favoured foodstuffs.

As noted, both Jackson and Parfitt consider the use of habitat primarily of fallow deer within the Forest Inclosures. It is, however, clear that many of the Forest fallow do forage out on to the Open Forest; deer near the Forest margins, or those whose range abuts agricultural land within the Forest perambulation, may also regularly graze out on to agricultural pasture, increasing their effective use of grassland areas. A more recent study (Thirgood 1990) attempted to determine patterns of habitat use from transect routes selected to sample more effectively all vegetation types available to the Forest deer in different areas within the Forest – inside and beyond the fenced Inclosures. Thirgood's observations were conducted over 2 years in each of five sites (Thirgood 1990, 1995a). Few differences were recorded between habitat use of male or 'small-deer' herds (females and followers) and the overall patterns of habitat use resolved were broadly similar to those described by Jackson and Parfitt (Table 3.6).

Overall, the habitat used most extensively in Thirgood's studies was again open woodland – with, over the year as a whole, extensive use made also of woodland clearings, grasslands and closed woodland areas. Significant seasonal variation in the degree of use of different habitats was recorded: the most pronounced changes were in the utilization of open woodland (highest in autumn and winter) and utilization of grasslands, which – as recorded earlier by both Jackson and Parfitt – was highest during spring and summer.

Table 3.6 Seasonal pattern of habitat use of New Forest fallow deer from observations of Thirgood (1990). Observations extend to include movements off the Forest on to adjacent agricultural or pasture lands.

	Winter	*Spring*	*Summer*	*Autumn*
Open deciduous woodland	43	37	27	34
Closed deciduous woodland	7	5	5	0
Open coniferous woodland	10	7	7	13
Closed coniferous woodland	5	0	0	20
Mixed woodland	0	5	5	8
Rides and glades	7	7	11	7
Grassland	12	25	25	0
Heathland	6	0	8	5
Others	10	14	12	13

While Thirgood's more recent analyses do not materially affect the conclusions reached earlier by Jackson or Parfitt, the extent of use of open grasslands (as opposed merely to grassland areas within woodlands: as rides or clearings) is more explicitly assessed and the overall use of grasslands of all types in summer is clearly slightly higher than that reported by the earlier studies in Inclosures only.

Patterns of use of different habitats in the five different areas of the Forest surveyed by Thirgood may be related to the actual area of each habitat available, to offer some measure of the degree to which animals were actively selecting, or avoiding, particular habitat types, rather than merely occupying them in direct proportion to their relative available area within their home range. Woodland clearings were positively selected in all sites and heathland actively avoided (used less than would be expected in relation to the area available). Open woodland was, as noted, the most extensively utilized habitat type in all sites – even though it was not equally available. Thirgood (1995a) cites as an example patterns of woodland use by female groups in two of his five areas. While the availability of open woodland differed between the two sites (58.1% and 36.2% of the sampled area) the animals in both areas utilized this habitat type to a similar extent (42.3% and 44.2% of all observations over the year as a whole); this was reflected in a change from weak avoidance to positive selection in the two sites.

Patterns of habitat use described by Jackson, Parfitt and Thirgood are all based on direct observation of animals from transect routes established within their range and walked on a regular basis. While transect routes were in all cases established in order to ensure a representative sample of all habitat-types considered – and in similar proportion to that in which they occur within the animals' range – all are to an extent hampered by differential visibility in different habitat types (although this is not such a great problem in the open-structured communities of the New Forest where all the understorey and shrub layer have been removed by centuries of grazing); more seriously, such visual observation is restricted to daylight hours. Fallow tend to be crepuscular, even nocturnal when in areas subject to any great level of disturbance, and may exhibit diel variation in habitat use. Although all these studies attempted to survey throughout daylight hours (with transects repeated around first light, and through the twilight into dusk, as well as during the middle of the day), we must accept that they present a consistent picture of habitat use patterns for daylight hours only. Some limited spotlighting revealed that the deer did make extensive nocturnal use of more open habitats (Thirgood 1990); but spotlight counts alone rarely give an accurate estimate of night-time patterns of habitat use, since dense habitats are invariably undersampled (McCullough 1982). No radiotelemetric studies of fallow deer in the New Forest have yet been reported.

3.4 DIET

The most comprehensive analysis available of the diet of New Forest fallow deer is that of Jackson (1974, 1977), whose results were based on the identification of plant remains in rumen samples of 325 deer killed during routine culling operations to control numbers (p. 24) or as the result of road traffic accident. His results (Table 3.7) show that through most of the year the deer are primarily grazers. Throughout the growing season from March to September, grasses form the principal food (comprising in the region of 60% of total food intake); herbs and broad-leaved browse also make a significant contribution. Acorns and mast are a characteristic food through autumn and early winter, although their importance in the diet varies from year to year with variations in the mast crop. Other major foods through the autumn and winter – on which the deer rely heavily when the year's mast crop is exhausted – are bramble, holly, ivy, heather and needles from conifers felled during forestry operations. Even at this time however, grass still makes up more than 20% of the diet. It is evident from this that (as might be predicted from the anatomical structure of the gut; Hofmann 1973, 1985) the deer are preferential grazers throughout the year, and take increasing amounts of browse over autumn and winter merely to compensate for lack of grasses and herbs outside the growing season (see also Caldwell *et al.* 1983).

Table 3.7 Diet of New Forest fallow deer (1970–73). Percentage composition of the main forage components in the diet of New Forest fallow deer, based on ruminal analyses of Jackson (1974, 1977).

	Jan.	*Feb.*	*Mar.*	*Apr.*	*May – July*	*Aug.*	*Sept.*	*Oct.*	*Nov.*	*Dec.*
Grasses	21	25	59	67	63	57	58	33	26	21
Forbs	1	1	1	6	6	12	7	2	2	1
Conifer browse	14	14	7	1	t	0	t	t	8	17
Holly	12	17	9	7	4	3	1	t	2	7
Other broad-leaves	1	1	t	5	14	11	6	4	12	4
Heather	16	24	16	3	4	3	2	1	8	16
Bramble/ rose	17	7	2	3	3	12	7	10	10	8
Ivy	8	4	1	1	4	1	2	0	4	4
Gorse	t	t	0	0	t	t	t	1	t	t
Mosses	1	2	1	1	1	t	t	t	1	t
Ferns	2	1	t	1	t	t	t	t	1	1
Fruits/mast	2	0	0	0	0	0	14	41	22	15
Others	5	4	4	5	1	1	3	9	4	6

t = Trace.

A subsequent analysis of diets of New Forest fallow deer by Parfitt, based on microscopical analysis of plant remains in faecal material, produced a remarkably similar picture of dietary composition and seasonal change. Despite the fact that it was based on different methods (the examination of faecal rather than ruminal materials) and represented the diet over a totally different time period, some 10 years after Jackson's study, basic results were strikingly similar to those of the earlier work, suggesting an identical pattern of forage use (Parfitt unpublished; summarized in Putman 1986b).

However, both studies produce a general picture of the diet of fallow deer through the Forest as a whole. Neither looked at possible differences in diet of deer in different areas within the Forest, or resolved in detail differences between males and females in diets selected. Based on ruminal analyses as well as direct feeding observations, though with small sample sizes for males, Jackson (1974) concluded that diets of bucks and does were very similar through autumn and winter and detected no significant separation. Yet differences between the sexes in both species composition and quality of the diet have been reported for a number of other dimorphic ungulate species. Thus Staines and Crisp (1978) reported differences in the quality of foods consumed by male and female red deer, and since that time similar differences have been noted by Takatsuki (1980) for sika deer, Shank (1982) for Rocky Mountain sheep, and McCullough (1985) in white-tailed deer (*Odocoileus virginianus*).

Putman *et al.* (1993) have more recently examined composition of the diets of male and female New Forest fallow deer, in a number of distinct situations: where sexual segregation is complete and bucks and does occupy non-overlapping home ranges, and in areas where the sexes, although socially segregated, occupy overlapping ranges (as Staines *et al.* 1982 for red deer). As in Parfitt's studies, the monthly composition of the diet of male and female fallow deer was determined by faecal analysis. Dung samples were collected opportunistically during the course of other observations; only fresh droppings were selected and were taken only when the sex of the animal depositing the pellet group was known (when a known individual was observed in the act of defaecation or by collection of samples following dispersal from a feeding site of a single-sexed group).

If samples from all sites are pooled, irrespective of context, merely to give an estimate of the dietary composition of male and female fallow deer within the Forest as a whole (Table 3.8), diets reported and patterns of seasonal change accord well with dietary profiles for the New Forest fallow deer presented by Jackson (1977) or Parfitt. These independent studies again reveal fallow deer in the New Forest as predominantly grazers. Grasses dominated the diet (making up nearly 70% of the diet) and were particularly heavily utilized over the summer period; sedges and rushes, moss and dwarf shrubs increased in importance during winter and spring (Putman *et al.* 1993). However, differences were found in diets of male and female deer, whether diets are compared of animals occupying different geographical ranges, or of the two sexes in sympatry. Allopatric does were shown to eat significantly more grass than bucks throughout the year, with less use made of rushes/sedges, dwarf shrubs and other browse materials. Dietary differences between the sexes were less pronounced in areas of sympatry, although bucks still took significantly higher proportions of coarser grasses and browse (Putman *et al.* 1993). Overall, bucks had higher-quality diets in autumn and winter, whereas does had higher-quality diets in spring and summer (Putman *et al.* 1993).

Differences in diet between animals of non-overlapping ranges might in part be attributable to differences in relative availability of the different forages in male and female ranges – but even if differences in diet are in part consequential, the fact that the sexes **are** choosing to occupy distinct geographical ranges, which thus results in a difference in dietary composition, is in its own way just as significant as are differences in diet reported between the sexes in sympatry, where resources available must be presumed to be equivalent. Further, differences in diets of the two sexes between sites (i.e. variation in the diets of males, or females, between sites where they occurred in sympatry and those where they were allopatric) were most significant for males, suggesting that differences are not simply due to variations in availability of foodstuffs, but may be imposed upon the males by the presence/absence of the does themselves.

Table 3.8 Diet of New Forest fallow deer (1987–89). Percentage composition of main forage components in the diet of male and female deer, based on the faecal analyses of Putman *et al.* (1993). Contribution to the diet of fruits and autumn mast cannot be assessed from faecal samples.

	Jan.	*Feb.*	*Mar.*	*Apr.*	*May*	*June*	*July*	*Aug.*	*Sept.*	*Oct.*	*Nov.*	*Dec.*
Male diet:												
Grasses	75	61	52	54	74	76	66	79	70	73	83	64
Forbs	0	0	1	t	t	2	t	t	t	2	0	1
Sedges/rushes	4	1	10	5	2	1	2	1	6	5	6	14
Conifer browse	0	0	0	0	0	0	0	0	0	0	0	0
Holly	1	17	12	7	5	4	4	6	2	1	2	1
Other broad-leaves	3	3	0	6	13	10	15	3	11	4	1	2
Heather	3	5	3	3	2	3	1	3	2	4	1	6
Bramble/ rose	0	0	0	0	0	0	0	0	0	0	0	0
Ivy	0	0	0	0	0	0	0	0	0	0	0	0
Gorse	0	0	0	0	0	0	0	0	0	0	0	0
Fruits/mast												
Mosses	13	9	13	21	2	3	8	8	4	11	7	12
Ferns	0	t	0	0	t	t	1	t	t	t	t	0
Others	1	4	9	4	2	1	3	0	5	0	0	0
Female diet:												
Grasses	70	64	65	64	65	85	82	78	75	71	58	46
Forbs	0	0	0	0	0	0	0	0	0	0	0	0
Sedges/rushes	5	3	8	8	2	1	4	6	8	11	7	13
Conifer browse	0	0	0	0	0	0	0	0	0	0	0	0
Holly	2	8	9	1	2	6	1	1	2	1	6	5
Other broad-leaves	1	6	0	1	22	5	5	7	5	5	2	5
Heather	6	4	4	2	2	1	2	2	6	2	10	12
Bramble/ rose	0	0	0	0	0	0	0	0	0	0	0	0
Ivy	0	0	0	0	0	0	0	0	0	0	0	0
Gorse	0	0	0	0	0	0	0	0	0	0	0	0
Fruits/mast												
Mosses	15	10	14	24	6	2	5	4	3	10	10	12
Ferns	0	t	t	t	t	1	t	1	t	t	2	4
Others	1	5	0	0	1	0	1	1	1	0	5	3

t = Trace.

Analyses reported by Putman *et al.* (1993) suggest significant variation in all seasons in composition and quality of the diet of fallow bucks and does, both where the two sexes occupy distinct geographical ranges and where they occur in sympatry. Differences in the diets recorded for animals from different sites in the current study were most significant for males, suggesting again that these differences are not simply due to variations in availability of foodstuffs, but that male diets are showing greater adjustment than females in accommodating overlap with the other sex. Such conclusions fit predictions from a model of dietary competition between the sexes based on the allometry of relationships between body mass and bite size, leading to the exclusion of larger-bodied males in species with distinct dimorphism from preferred swards (Clutton Brock *et al.* 1987, Illius and Gordon 1987).

Plate 1 The New Forest is not just a forest of trees; only 20% of the area is covered by ancient woodland.

Plate 2 Wet or humid heathland develops where drainage is impeded.

Plate 3 Dry *Calluna* heathlands occur on the better-drained plateau gravels of the Open Forest.

Plate 4 Acid grassland merges into the richer grazing of streamside lawns on the alluvial soils of the river floodplains.

Plate 5 In an attempt to increase the grazing areas available to cattle and ponies some of the former acid grasslands have been ploughed, fertilized and reseeded.

Plate 6 The different vegetation types of the Forest are patchworked together into an intimate mosaic.

Plate 7 Due to the heavy grazing pressure they sustain, the Forest grasslands boast little vegetation taller than a few millimetres.

Plate 8 All woody browse is removed up to a height of 1.8 m, see plate 9.

Plate 9 A clear browse line develops in all the Forest woodlands, see also plate 8.

Plate 10 Ponies which spend a high proportion of their time feeding out in open heathland throughout the year were generally found to be in poorer body condition.

Plate 11 Fallow are the most numerous of the Forest's wild deer.

Plate 12 A group of fallow does and fawns forages at the woodland edge.

Plate 13 Red deer have a very local distribution within the Forest and have never occurred in large numbers.

Plate 14 Sika deer established themselves in the Forest at the very beginning of this century.

Plate 15 Roe deer have declined in numbers within the New Forest over recent years.

Plate 16 Performance of roe has declined as habitat structure within the Forest has changed, with greatly reduced fawning rates.

4

Behaviour and ecology of sika, red and roe

4.1 BEHAVIOUR AND ECOLOGY OF NEW FOREST SIKA

Sika deer (*Cervus nippon*), like fallow, owe much of their current global distribution to human introduction. Native to the Japanese islands and adjacent mainland of Asia, sika have been introduced successfully to much of central Europe (including Austria, the Czech Republic, France, Germany, Poland) as well as to New Zealand and a number of States within the USA (Maryland, Virginia, Texas). Where they have established successful feral populations, they have become widely recognized as a serious pest of forestry; hybridization with red deer has proved a problem in Eire and the United Kingdom (e.g. Harrington 1982, Ratcliffe *et al.* 1992, Putman and Hunt 1994). In addition, it is notable that in many areas of introduction, competition has been reported between sika and other species of native (or introduced) wildlife (e.g. Kiddie 1962, Feldhamer *et al.* 1978, Challies 1985, Keiper 1985, Feldhamer and Armstrong 1993). Sika were introduced into the UK at around the turn of the century: with the first recorded introduction dated around 1860 and numerous further introductions documented up until the 1930s. Ratcliffe (1987a) offers a detailed review of the history of introduction and the subsequent spread of the species through Great Britain; numbers in the UK were estimated by Harris *et al.* (1995) at 11 500. New Forest populations derive from escapes from a private collection in the nearby Manor of Beaulieu, where they were first introduced in 1904.

The taxonomy of sika deer has been revised many times, but there is a general consensus that we may recognize two distinct types: the smaller sika deer of the Japanese islands (*C. n. nippon* and related forms) and the mainland Asiatic sika (*C. n. hortulorum* and its allies) consisting of the larger Manchurian and Formosan forms (Ratcliffe 1987a). Animals of both taxonomic types have been introduced into Britain (in the first introduction of 1860 for example, the Zoological Society of London was presented with specimens reported as *C. n. nippon* and *C. n. hortulorum* and further introductions of both putative subspecies have been documented since); however, it is believed that most of the sika held in British parks or collections were of the island form, and certainly the small, Japanese sika is the only type known to have become established in the wild (Lowe and Gardiner 1975).

4.2 SOCIAL ORGANIZATION

In an analysis of over 3000 observations, Mann (1983) found a pronounced annual cycle of group size in New Forest sika (Figure 4.1), similar to that discussed earlier amongst fallow deer. Once again the basic social 'unit' is the individual (male, female or female with calf), but among sika this is far more obvious; indeed sika are one of the less social of the deer species (Putman 1981, 1993) and through much of the year the individuals remain completely solitary. From the end of winter through until September, the animals are generally encountered alone, or as hinds accompanied by a calf. The rut in September causes an increase in group size and increases the number of groups encountered of mixed sex; rich autumn food supplies allow these larger aggregations to persist through until March or April, when females drift away from the groups to calve (Putman and Mann 1990). Even during this winter period, however, sika never establish groups as extensive as the feeding aggregations that may be formed by fallow. Indeed, they are rarely observed in groups of more than five or six; the largest group ever recorded in Europe consisted only of 12 individuals.

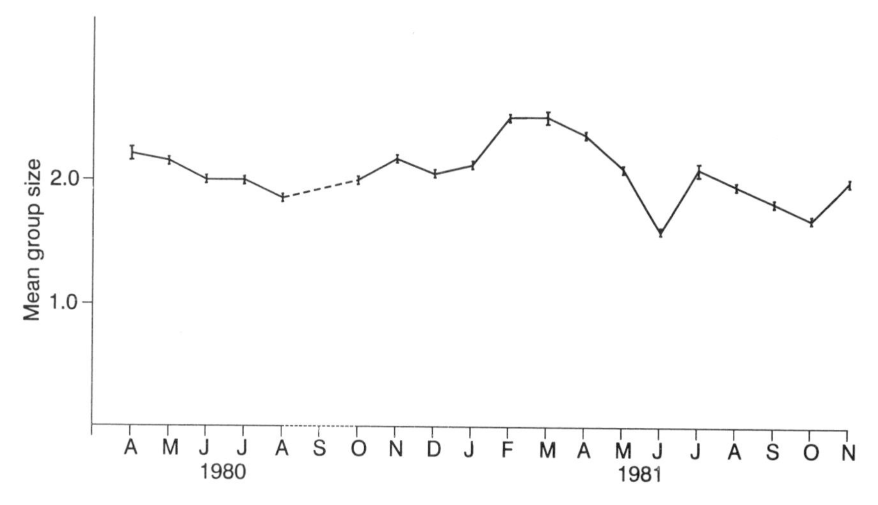

Figure 4.1 Seasonal changes in the group size of New Forest sika deer (from Mann 1983, with permission).

Sexes are strongly segregated throughout the year, and within the New Forest until the early 1980s, stags and hinds occupied distinct geographical ranges. More recently, as densities have increased and the animals disperse more widely through the Forest, this geographical separation is less pronounced (S. Smith, personal communication), although social segregation between the sexes is maintained. Early records of the rutting behaviour of sika deer in Britain (Horwood and Masters 1970) suggest that stags mark out and defend mating territories in woodland within the hind range. These territories are marked by thrashed *Calluna* bushes and other low vegetation and frayed perimeter trees. But it is clear that mating strategy within sika is every bit as flexible and responsive to environmental circumstance as that of fallow. In various different populations now studied in Britain and elsewhere in mainland Europe, males have been found to defend rutting territories, as above, collect and defend a harem, as do red deer, or merely wander throughout the female range in search of oestrus hinds (Putman 1993). The development of a simple breeding 'lek' has even been reported in certain populations (Bartos *et al.* 1992, Putman 1993).

4.3 HABITAT USE

Sika deer are generally regarded as characteristic of somewhat acid soils. 'Typical' sika habitat in Britain could be considered a mix of heathland and coniferous woodland (two vegetation types that are often associated anyway, in that many commercial forestry plantations are established on areas of native heath). However, the animal is clearly an opportunist and can readily adapt to a wide range of conditions (Takatsuki 1987, Mann and Putman 1989a,b).

In the 'typical' British habitat of acid coniferous woodland, sika show a very predictable pattern of habitat use (Horwood and Masters 1970, Mann 1983). Relatively little forage of high quality is available to the deer within the forest plantations; when feeding, the deer tend to leave the cover of the woodland for the open vegetation beyond, in heathland or on agricultural fields. Use of such open habitats is primarily at night (the deer are very sensitive to human disturbance and thus tend to venture out into more exposed habitats only after dark; cf. van de Veen 1979, Langbein and Putman 1992); by day the animals retire to the woodlands and lie up in dense cover. This same shyness of any disturbance causes the deer to seek out the densest thickets when resting. Pattern of use of the range is thus in general extremely simple, with the animals lying up in these dense thicket areas during the day and moving out on to the adjacent heaths or agricultural land to feed at night. This regular daily migration is maintained throughout the year; indeed the overall pattern of use of available habitats changes little between the seasons (Mann 1983, Mann and Putman 1989a).

By comparison to the coniferous forest that they occupy over much of their range, the New Forest offers the deer a much more varied environment – and one of predominantly deciduous woodland. As noted in Chapter 2, the distribution of sika within the New Forest was, until comparatively recently, curiously restricted, with the population apparently confined to the one relatively small area in the south first colonized in the early 1900s. The core of this area is wooded: mostly mature oak–beech woodland with a secondary canopy of holly, yew and, unusually in the New Forest, both hawthorn and blackthorn. There are extensive patches of birch scrub, and within the area various blocks have been cleared and planted with conifers. The entire area is broken up with many clearings and rides, and the woodlands are surrounded by extensive tracts of heath and small areas of agricultural land. With this more varied habitat on offer, sika deer within the New Forest show a much more variable pattern of habitat use and show more pronounced seasonal variation in communities exploited than in other, less heterogeneous environments (Mann and Putman 1989a).

Table 4.1 Habitat use of New Forest sika from data of Mann (1981–82); in Mann and Putman (1989a). Values shown are the percentage of all observations in any month recorded in each habitat.

	Jan.	*Feb.*	*Mar.*	*Apr.*	*May*	*June*	*July*	*Aug.*	*Sept.*	*Oct.*	*Nov.*	*Dec.*
Deciduous woodland	50	52	50	55	48	31	37	32	35	55	74	62
Coniferous woodland (mature)	0	0	0	0	0	0	0	0	0	0	0	0
Coniferous woodland (thicket/pole)	20	17	18	16	16	23	20	20	12	18	10	11
Plantation and prethicket	6	4	11	16	18	17	23	21	26	7	3	6
Rides and glades	24	27	26	13	23	29	30	27	26	18	13	21
Acid grassland/ wet heath	0	0	0	0	0	0	0	0	0	0	0	0
Dry heath	0	0	0	0	0	0	0	0	1	1	0	0
Bog	0	0	0	0	0	0	0	0	0	0	0	0

Analyses of habitat use (Table 4.1) were based on direct observations from regularly traversed transect routes sampling much of the core area of the sika's initial distribution within the New Forest (Frame Heath Inclosure). Observations in this case extended throughout the 24-hour period and the pattern of the results is substantiated by independent analyses of rates of faecal accumulation in the different habitats (Mann

1983). Over the winter period, the animals were found to spend the majority of their time within areas of oak–beech woodland, continuing to exploit these mixed woodlands at night during the spring and early summer, but increasingly making extensive use of a variety of other habitats, during daylight hours – particularly areas of young conifer (prethicket). During the summer, the majority of animals fed during the day in prethicket areas (although all habitats are used to a certain extent), benefiting from the security and cover provided by the relatively dense conifers, but exploiting the comparatively abundant forage available between the trees at this stage, before canopy closure. Diet at this time of year (below, p. 54) shows a high intake of grass and broad-leaved browse, and these are readily available in all habitats. In the autumn, when acorns and leaves fall, the resources offered by the various different habitats, in terms of both food and thermal cover, change rapidly. Prethicket areas of conifer before canopy closure still provide a good food supply, but this is now inferior to that offered by the tree fruit falling in the oakwoods – which also provide a considerable degree of shelter. In late autumn and winter, when the acorn crop has been exhausted and most of the fallen leaves are decaying, food supplies within the Forest become more limited; at this time the deer were found to start to feed more extensively on coniferous browse and *Calluna*. These are available in a variety of habitats; the animals still spent most of their time within the oakwoods, but were also recorded foraging out into prethicket and heathland areas (Table 4.1; Mann and Putman 1989a).

The data presented above derive from studies carried out on animals within the area of the sika's initial colonization of the New Forest (Frame Heath), and before the animals had begun to spread beyond this one highly localized area (Mann 1983, Mann and Putman 1989a); it is in addition an area where the sika were specifically favoured by management. Red deer do not occur in the area, roe are scarce and fallow deer were systematically culled on sight in an attempt to safeguard the sika. In effect these earlier studies were thus of sika deer geographically isolated from other Forest herbivores.

However, since that time sika have begun to disperse more widely through the Forest, colonizing a wider range of environmental types and spreading into areas where they co-occur with others of the Forest's ungulate species. A more recent study by Michael Boxall (1990) offers us a comparable analysis of patterns of habitat use by sika from another site within that wider range. Roydon Wood (380 ha) is, strictly speaking, outside the statutory Forest of today (although it was included within the perambulation until comparatively recently). However, it lies immediately adjacent, abutting the boundary of the 'official' Forest to the south. Habitat composition of the study site is comparable to that of Mann's New Forest study area. Much of the area is mixed broad-leaved woodland (predominantly oak, with birch and beech); there are two substantial blocks of Scots pine and several smaller areas of pine and European larch. There are also extensive areas of both dry, *Calluna*-dominated heathland and wetter areas dominated by *Erica tetralix* and *Molinia caerulea*. Approximately 65 ha of the area is abandoned agricultural land or unimproved pasture, and the site is bounded by largely arable agricultural land.

Boxall's data are based on visual observations of animals from regular fixed transect routes over a 4-year period from 1985 to 1988. While sika in Roydon Woods, like those in the Frame Heath area of the Forest, make considerable use of the areas of mixed broad-leaved woodland throughout the year, they did not use them as extensively at any time (Table 4.2). They made far greater use of the available areas of wet heath and, during winter and spring in particular, the mature pine stands. In winter, the animals made greatest use of areas of wet heath, broad-leaved woodland and the mature conifer stands; through late spring and summer use of wet heath and pine forest declined to some extent as increasing use was made of the open grasslands beyond the woodland margin, and woodland rides and glades. In autumn, stags began to use the pastures less, drifting back into the woodlands and increasing their use of heathland communities to winter levels; hinds continued to graze in the open pastures longer before they, too, moved back to feed in the mature pinewoods and heathlands.

Table 4.2 Seasonal pattern of habitat use of Roydon sika from observations of Boxall (1990). Figures show percentage of all observations recorded in each habitat.

	Winter	*Spring*	*Summer*	*Autumn*
Mature deciduous woodland	33	28	31	25
Coniferous woodland	21	25	11	15
Rides and glades	8	7	15	8
Wet heath	33	27	18	28
Dry heath	4	1	2	7
Grassland/ arable land	1	13	23	17

4.4 DIET

Once again we may contrast the diet of sika deer within the New Forest with that observed elsewhere within their British distribution. Mann (1983) used both ruminal and faecal analyses to tease out a picture of the foraging ecology of sika deer in southern Britain. The complete annual diet of sika in Wareham Forest (a commercial coniferous forest in Dorset) and the New Forest was determined through identification of plant cuticular remains in fresh faecal pellets; information on winter diet was supplemented by analysis of rumen samples from animals shot during routine culling operations.

At Wareham, the diet was shown to be relatively constant throughout the year, with a high intake of grass (30–40%) and *Calluna* (40–50%) in all seasons; a variety of other dietary components contribute to the remainder of the diet (pine needles, bark, gorse) but no single item comprised more than about 8% at any time. Grasses consumed were principally *Molinia caerulea* (50% of all grass taken), *Agrostis curtisii* and *A. capillaris* (Mann 1983, Mann and Putman 1989b).

This dietary profile – a diet composed principally of grasses and heather, and seasonally unchanging – is not peculiar to this one site, but is in fact rather generally characteristic of mainland British sika – although, if anything, animals in other populations studied (Mann and Putman 1989b) consumed rather less heather and more grass, with grasses sometimes comprising 70% or more of the total diet. A recent study of the diet of sika deer in Eire (Quirke 1991) offers similar conclusions. Killarney sika of Quirke's study made up some 60% or more of their diet from grasses and other monocotyledons (winter/spring, 39%; summer/autumn, 77%) and *c.* 15% of ericaceous shrubs (winter/spring, 21.3%; summer/autumn, 7.3%). In general then, sika deer in Britain may be regarded primarily as grazers, a result compatible with the limited data available for their diet in their native range. Prisyazhynuk and Prisyazhynuk (1974) writing of sika deer on Askold Island (in the former USSR), report that the bulk of the food taken is grasses; Furubayashi and Maruyama (1977) found that sika in the Tanzanawa Mountains in Japan consume 106 different plant species but feed primarily as grazers. Takatsuki (1980) used faecal analysis and direct examination of feeding sites to investigate the diet of sika deer on Kinkazan – another of the Japanese islands – and again concluded that the major part of the diet year round, was grasses.

By comparison, the diet of the New Forest sika appears somewhat unusual (Mann and Putman 1989b), for New Forest sika take considerable quantities of both deciduous and coniferous browse, particularly in the winter, when it may comprise up to 23% of the total food intake. In addition, the animals show striking seasonality in diet, feeding very opportunistically on foods as they become available (Table 4.3).

In spring and summer, New Forest sika do feed extensively on grasses (about 30% of the diet during spring, up to 40% in summer) and *Calluna* (30% in spring; 35% in summer) as do populations elsewhere in Britain, but their diet is far more varied and includes significant amounts of other

Table 4.3 Diet of New Forest sika deer. Percentage contribution of different forages to the diet of New Forest sika (data from Mann 1983, after Mann and Putman 1989b).

	Jan.	*Feb.*	*Mar.*	*Apr.*	*May*	*June*	*July*	*Aug.*	*Sept.*	*Oct.*	*Nov.*	*Dec.*
Grasses	25	25	22	39	38	40	39	50	44	31	28	27
Forbs	0	0	0	0	0	0	0	0	0	0	0	0
Conifer browse	20	19	23	13	2	0	0	1	0	6	8	16
Holly	1	3	1	1	1	1	1	1	2	2	1	1
Other broad-leaves	11	11	10	13	14	14	16	10	19	25	14	14
Heather	24	23	30	23	35	37	35	29	27	23	24	25
Bramble/rose	0	0	0	0	0	0	0	0	0	0	0	0
Ivy	0	0	0	0	0	0	0	0	0	0	0	0
Gorse	7	14	8	7	6	5	6	6	7	4	7	5
Fruits/mast	6	4	2	0	0	0	0	0	0	6	14	9
Others	6	1	4	4	4	3	3	3	1	3	4	3

forages: deciduous browse, gorse and conifer needles. In autumn, only 25% of the diet is composed of heather and grass, with the bulk of the food intake being composed of pine needles (from coniferous browse), gorse, holly and acorns. In winter, there is a further increase in the intake of pine needles, and as might be expected a decline in the intake of deciduous browse and forbs; at this time less than 20% of the diet is made up of grasses. Such a heavy emphasis on browse throughout the year is unusual and cannot easily be explained in terms of changing forage availability. Mann (1983) suggests that in the New Forest, sika deer may be forced by competition with other herbivores to take larger quantities of browse material. Although, as noted, roe deer are uncommon in that part of the Forest where the sika occur, and fallow are deliberately shot out from the sika's stronghold, horses are abundant. New Forest ponies and other domestic animals turned out on to the Forest grazing, are officially excluded from the Forestry Inclosures; in practice they are free-ranging and it is impossible to keep them out of even fenced Inclosures. In summer, forage productivity is probably sufficient to support offtake; in winter, however, food resources become more limited. The ponies, by virtue of their opposed incisor teeth, can crop closer to the ground than the deer – and as their monogastric digestion relies on a rapid throughput of large volumes of material, it might be expected that large amounts of the limited grass of Forest rides and glades will be consumed rapidly. It seems possible that the deer cannot compete effectively at this season and this may be why they feed so extensively on coniferous browse, a resource not exploited by sika in other areas (Mann 1983).

4.5 RED DEER

Although presumably native to the area, numbers of red deer in the New Forest over the past 200 years have probably never exceeded 80–100 animals. It is probable that they were even extinct within the Forest until re-establishment in the early 1960s from English park and Continental stock. Until comparatively recently, numbers remained extremely low and populations were restricted, rather like those of the sika deer, to very localized distribution in two distinct areas in the south of the Forest. More recently numbers have been increasing and the deer have begun to colonize much larger areas of the Forest; the two separate 'sub'-populations are now effectively contiguous and red deer may be encountered virtually throughout the southern part of the Forest. Because this expansion in both range and population number is comparatively recent, thus far relatively little research has been undertaken on the ecology or behaviour of the Forest reds. However, while as yet there is no detailed published analysis of diet, we do have some extensive information on habitat use and some limited (observational) data on forage selection from the continuing studies of David Payne and Graham Long (Payne 1987).

Sexual segregation among the red deer of the New Forest is not particularly pronounced. Adult males and females occupy the same geographical areas within the Forest year round (although additional mature males from outside the Forest do come into the area during the autumn rut) and of some 378 sightings of groups or individual red deer over a 10-year period, only 10.3% were of groups containing only males (and even these were usually solitary individuals); groups containing hinds and followers comprised 34.4% of all groups observed, while 55.3% of groups observed contained adults of both sexes.

Group size and composition varied with season. Groups of males only were mainly small, with more than 90% of all observations being of groups of one or two individuals. Female groups were also characteristically small; although aggregations of up to 30 individuals were recorded on occasion, 66.9% of records were of groups of four or fewer and groups were characteristically smaller during summer and autumn (June–October) than at other times of year. Mean size of mixed-sex groups throughout the year was between 10 and 11 individuals (Payne 1987).

Data on habitat use derive from direct observations recorded during a single year (December 1985–November 1986) in a number of different sites (Payne 1987). Throughout the year, red deer were most commonly observed in areas of wet and dry heathland – out on the open heath, or where the heathland had become invaded by self-set pine; such observations accounted for more than 65% of all records throughout the year.

Conifer thickets and mature stands of Scots pine were increasingly used through autumn and winter, and areas of mature deciduous woodland were exploited in autumn at the time of the mast fall. Animals were seen to use areas of grassland (whether improved areas or natural acid grassland) during the spring and summer only; as the summer progressed and these communities dried out and became sere, use of bogs and wet flushes increased significantly (Table 4.4). Payne (1987 and personal communication) points out that these data may be biased towards higher reported use of heathland, both as a result of differential visibility in different habitats and also through visiting areas of heathland more frequently).

Table 4.4 Seasonal pattern of habitat use of New Forest red deer, recorded by Payne (1987); figures show percentage of observations in any season recorded in each habitat, December 1985 – November 1986.

	Winter	*Spring*	*Summer*	*Autumn*
Mature deciduous woodland	0	0	2	11
Coniferous woodland (mature)	18	0	3	3
Coniferous woodland (thicket/pole)	1	0	0	5
Plantation	9	2	3	7
Wet heath	15	22	10	11
Dry heath	51	42	50	53
Bogs	5	0	16	2
Grassland	0	29	16	6
Others	1	5	0	2
N =	212	153	127	627

Payne (1987) also records observations made on red deer feeding in different habitats. These observations of feeding behaviour were not made in any sense systematically, but subsequent analysis of faecal pellets (Long, unpublished data) has in some measure corroborated the overall picture presented. Heathers were the principal items noted for the year overall and accounted for 38% of feeding records. *Molinia* was also recorded extensively (22% of all feeding observations), with coniferous browse (14%), deciduous shrub (10%), rushes (5%) and grazing on short 'greens' (8%) making up the bulk of the remaining observations. Although heathers remained important throughout the year, other items varied in importance seasonally, reflecting the seasonal variation in habitat selection. Of winter feeding observations, 53% were recorded on heather, with conifer browse comprising a further 44%. *Molinia* accounted for 23–30% of all feeding observations recorded through spring, summer and autumn; in spring the deer were observed feeding on the short grass of improved grasslands (24% of all observations), but foraging on these grasslands decreased in summer (6%) as use of rushes and other wetland plants in boggy areas increased. Deciduous browse was taken most during the autumn (13% of observations).

It may be noted that the diet suggested from these preliminary observations appears, like that of the New Forest sika deer, to differ somewhat from that reported for the species elsewhere in Britain (Sharma 1994). Scottish red deer consume much greater amounts of broad-leaved grasses (bents and fescues), which may comprise over 90% of intake in some areas, during spring and summer (Clutton-Brock and Albon 1989). Although Highland red deer take more heather in eastern areas, where it is more abundant, than in the western Highlands (Clutton-Brock and Albon 1989), nowhere do they take *Molinia* in the quantity suggested by Payne's observations in the New Forest (Sharma 1994).

4.6 BEHAVIOUR AND ECOLOGY OF THE FOREST ROE DEER

Following their extinction throughout England during the sixteenth or seventeenth centuries, roe recolonized the New Forest area from about 1870 onwards, spreading eastwards from populations re-established in Dorset (Prior 1968, Jackson 1980). Census figures suggest a population of between 400 and 500 animals in the early 1970s, but since that time peak numbers have steadily declined (Putman and Sharma 1987; Table 2.1). Roe are now sparsely and patchily distributed within the Forest; as elsewhere within their range (Kurt 1978, Putman 1988), the Forest roe are seasonally territorial and largely solitary, although fawns may remain with their mothers until dispersal in late winter and small groups may be recorded over winter on favoured feeding grounds.

The smallest of the native European species, the roe is effectively the 'r'-strategist amongst the deer: an opportunistic species able to colonize a wide variety of different environmental types from dense woodlands to open agricultural prairies with little or no cover (e.g. Zejda 1978, Turner 1979, Kaluzinski 1982) and in suitable conditions capable of extremely rapid recruitment (e.g. Ratcliffe and Mayle 1992, Hewison 1996, Putman *et al.* 1996). (An excellent recent review of the general biology of roe is offered by Danilkin and Hewison (1996) in a companion volume in this series.) However, as a concentrate selector (*sensu* Hofmann 1985) with a feeding strategy dependent on selection of small morsels of highly nutritious foodstuffs, the roe is primarily a pioneer species, associated particularly with early successional communities, characterized by rapid growth and high production of the 'concentrated' foods that the roe favour; hence roe are themselves at their most productive in young woodlands or other disturbed habitats which offer a rich ground vegetation and shrub layer. Current distribution within the New Forest reflects this: roe occur at any density only in those areas of the Forest where open grazing is still available among young plantations, or where windthrow has opened clearings with abundant regeneration (Sharma 1994).

4.7 DIET

Any explanation of the distribution and performance of the New Forest roe population is dependent on an understanding of the very specific foraging requirements of this species. Small body size (18–28 kg), restricting overall gut capacity, and relatively poor ruminal development restrict their potential to digest highly fibrous diets and constrain them to a dependence on cell-solubles. Roe have long been recognized as highly selective feeders who pluck small and highly nutritious morsels from a wide variety of plant species: concentrate selectors in the terminology of Hofmann (1985). Although it is often presumed that such a feeding style necessarily equates with browsing habit, the assumption is unfounded; close examination of all published data shows that roe wherever they occur may take considerable quantities of grasses and forbs when these are at their most nutritious.

Their diet within the New Forest was described by Jackson (1980) from the analysis of 105 rumen samples collected between November 1970 and March 1973 from animals which had been culled in routine control operations, killed in road traffic accidents, or otherwise died within the Forest. Jackson's results (Table 4.5) showed that for New Forest roe, browse materials such as bramble and rose, the new growth of deciduous woodland trees and dwarf shrubs, formed the bulk of the diet throughout the year. Bramble and rose together formed between 25% and 45% of the diet throughout and comprised the largest food fraction in all months except January and April. From January to March, foliage from felled conifers or from young plantations, *Calluna* and ivy were major foods, with lesser amounts of grasses, herbs and fungi. Over the summer (growing) season, herbs and grasses became more important and new growth of deciduous trees and shrubs was also favoured. During the autumn acorns were a characteristically important element of the diet, when available; fruit and nuts formed up to 15% of the diet from September through till midwinter.

Although deciduous browse was present in the rumina all through the year, the amounts present from October to March were minimal; from April to September, however, such browse formed between 10% and 30%

Table 4.5 Diet of New Forest roe deer (1970–73) from ruminal analyses of Jackson (1980); figures show percentage contribution to the diet of the different forage types consumed.

	Jan.	*Feb/Mar.*	*Apr.*	*May*	*June*	*July/Aug.*	*Sept.*	*Oct.*	*Nov.*	*Dec.*
Grasses	4	5	10	7	8	8	8	9	10	4
Forbs	5	2	30	13	16	16	17	4	4	6
Conifer browse	33	22	5	1	8	0	0	12	12	13
Holly	0	2	0	0	14	1	0	0	0	0
Other broad-leaves	2	4	13	30	14	15	14	5	5	2
Heather	6	14	14	5	4	7	5	5	4	7
Bramble/rose	31	26	20	35	32	40	38	38	37	46
Ivy	12	22	7	6	4	3	3	11	11	2
Gorse	0	0	0	0	0	0	0	0	0	0
Fruits/mast	1	1	0	0	0	0	8	8	17	7
Mosses	0	1	0	0	0	0	0	0	0	0
Ferns	2	1	1	3	t	2	1	t	t	3
Others	4	0	2	0	0	8	6	8	0	10

t = Trace.

of the diet. Browse species consumed at this time included birch, beech, oak, hawthorn, willow and buckthorn, as well as the ubiquitous holly and shoots of bilberry. Twigs and foliage of coniferous species (primarily Scots pine and Norway spruce) were taken throughout the year, but were most important from October to March. Grasses consumed included *Agrostis capillaris, A. canina, Deschampsia flexuosa, Poa annua, Holcus lanatus* and *Sieglingia decumbens*. Roe deer seemed to avoid *Agrostis curtisii, Molinia caerulea, Deschampsia cespitosa* and *Brachypodium sylvaticum* (making interesting comparison with New Forest ponies, p. 78, and sika deer, p. 53).

Diet of roe in the New Forest, as described by Jackson, is very similar in general 'shape' to the diet found by other workers elsewhere in Great Britain or continental Europe (e.g. Hosey 1974, 1981, Johnson 1984). Because food availability and precise species composition varies so much between different localities, detailed comparisons of actual species eaten or precise composition of diets from different areas are of limited value; perhaps the most notable difference in diet between the New Forest population and those elsewhere in Britain is that ivy figured substantially in the winter diet only in the New Forest (Jackson 1980).

Jackson's data (see Table 4.5) were obtained during the early 1970s, when the roe populations within the New Forest were at a reasonably high density; the vegetational structure of the Forest has changed appreciably since that time – at least in terms of the availability of the kinds of early successional, early growth stage communities that provide the sort of forage resources that roe favour. As part of a longer-term study of factors associated with the recent decline in roe deer populations within the New Forest, Sharma (1994) has re-examined the composition of the diet of Forest roe from samples collected during 1989–90.

Over this period few roe, if any, were culled within the Forest. Sharma's analyses thus depended on examination of plant residues in faecal material; in addition, sample sizes were small – one of the inevitable hazards of working with a relatively rare species in any context! Microhistological examination of faecal materials collected from six different sites within the Forest suggested that the diet of roe over this period was strongly dependent on coniferous and broad-leaved browse (Table 4.6); each contributed some 25% of the annual diet overall. Other important food categories (contributing between 5% and 11% of annual diet) included grasses, forbs, ferns, heathers (mostly *Calluna vulgaris*) and bramble (Sharma 1994).

Table 4.6 Diet of New Forest roe deer (1989–90) from faecal analyses of Sharma (1994). Figures show percentage contribution to the diet of different forages consumed; contribution of fruit/autumn mast cannot be assessed from faecal samples.

	Jan.	*Feb.*	*Mar.*	*Apr.*	*May*	*June*	*July*	*Aug.*	*Sept.*	*Oct.*	*Nov.*	*Dec.*
Grasses	8	8	3	7	8	16	11	21	19	17	6	9
Forbs	6	14	10	16	14	8	8	13	13	8	4	6
Conifer browse	56	52	59	41	30	16	14	7	6	18	25	42
Holly	2	2	3	1	t	1	t	t	0	0	0	1
Other broad-leaves	12	6	8	14	35	39	48	25	32	26	28	27
Heather	4	7	6	12	4	8	3	6	7	3	8	3
Bramble/rose	2	1	2	2	5	6	11	9	6	5	3	1
Ivy	1	5	3	1	1	t	0	2	0	1	3	3
Gorse	0	0	0	0	0	0	0	0	0	0	0	0
Fruits/mast												
Mosses	4	2	1	1	1	t	1	1	2	8	2	5
Ferns	5	3	5	4	2	6	5	15	15	15	18	2
Others	0	0	0	1	0	0	0	1	0	0	3	1

There are some very clear differences between Sharma's data for 1989–90 and those for 1970–73 reported by Jackson (1980). Jackson, as noted above, found bramble, with rose, to be by far the most important item in the diet, accounting for some 40% of annual intake. (Similar, or even higher, importance of bramble has been noted in other studies in the south of England: Hosey 1981, Diakite 1983). By 1989–90, the proportion of bramble in the diet of New Forest roe had fallen to only 5%. By contrast, levels of intake of both deciduous and coniferous browse, grasses and ferns are far higher in Sharma's analyses. Broad-leaved browse and pine needles were found to be consumed over a greater period in the year than that recorded by Jackson, and were taken in larger relative quantity; grasses were taken in both studies throughout the year, but levels consumed over late summer reach double that reported by Jackson. Finally, ferns appear as an important item in the diet only in the 1989–90 analysis: contributing in Sharma's study some 8% of the overall measured diet (compared to the 1% reported by Jackson 1980).

To some extent, the differences noted here may be explained by differences in methodology. Jackson's (1980) analyses were based on point-frame sampling of rumen contents (after Chamrad and Box 1964) while Sharma's data derive from faecal analyses (after Storr 1961, Stewart 1967). Dietary estimation from faecal materials has several potential sources of bias – notably due to problems of differential digestibility and fragmentation (Putman 1984). Fibrous and cellulose-rich materials, including both evergreen and deciduous trees and shrubs, will tend to be digested less completely than softer forb material, which may disappear completely leaving no identifiable cuticular residue in the dung. In consequence, soft items are likely to be underestimated, and, as a result, the more fibrous materials that do persist are consequently overestimated in their proportional contribution to the diet overall. Dietary estimation from ruminal analyses has its own sources of bias. Less digestible materials which have a longer period of retention within the rumen will tend to be over-represented; ruminal analyses will thus again tend to overestimate the contribution of browse materials within the diet. In effect, however, the bias of both methods is actually in the same direction: overestimation (if not necessarily to the same degree) of more fibrous materials such as coniferous browse, heather or other dwarf shrubs. And differences noted by Sharma in the diets of roe deer in 1989–90, by comparison to those of the early 1970s, relate to fairly gross changes in the proportions of this less digestible fraction, as well as to changes in the proportion of forages less likely to be falsely represented, such as bramble or fern.

It seems likely therefore that the differences in diets recorded by Jackson and Sharma reflect a genuine change in diet of roe deer since Jackson's studies of the early 1970s, as a result of real changes in the availability of forage species, or their selection by roe (Sharma 1994). Dietary profiles reported by Sharma are certainly consistent with current patterns of habitat use.

4.8 PATTERNS OF HABITAT USE

Sharma's studies of the ecology of New Forest roe during the late 1980s also included a detailed analysis of patterns of habitat use of roe within those areas of the Forest where they still persist. Interested in the reasons for the observed recent decline in roe deer density and performance, he deliberately selected for observation a range of (six) sites where roe were at low or moderately low abundance and areas where they occurred at (comparatively!) higher density.

Patterns of habitat use resolved derive in the main from direct observation from transects established in each study site and chosen to sample all habitats included within the animals' range in proportion to their relative abundance. Because of the necessary paucity of observation of animals in areas where they are known to occur at very low density, results from individual months are pooled to present a seasonal overview of habitat use (Table 4.7). Roe in Sharma's studies made high use of the three habitat types most extensive in their cover within his study areas. Thus approximately a third of all observations were recorded in open-based mature conifer stands with no understorey, in mature conifer stands with an understorey containing forage species known to be selected by roe (Tables 4.5, 4.6), and in broad-leaved, deciduous woodland with food-bearing understorey. Moderate use was also made of prethicket and establishment (plantation) areas of pine or other conifers – areas that offer forage and cover – and of deciduous woodland with a depleted field layer (Sharma 1994). Patterns of use, and the actual availability of the different habitats differed very significantly: roe were clearly not using the habitat types simply in proportion to their relative availability in the environment, but were positively selecting, or avoiding, different vegetational types. All those habitats considered to contain potential forage were strongly selected; broad-leaved woodlands with poor field and shrub layers were also selected, but to a lesser extent than those with more abundant field layer. Food-depleted conifer stands were avoided in every season, while coniferous blocks providing some degree of forage were selected from winter through to summer. Rides were selected strongly in summer and avoided at other times, but the broad-leaved woodlands dominate use overall and are selected very strongly in autumn, whether food-depleted or food-bearing (Sharma 1994).

4.9 ROE DEER HABITAT USE AND POPULATION PERFORMANCE

Sharma's studies of habitat use by roe embraced a number of different sites within the Forest where roe are considered to be at reasonable density/performance and others where numbers and performance have recently declined. There were clear differences between these sites in the overall availability of different habitats, with poor-performance sites containing substantially more coniferous woodland (mature and prethicket) devoid of any food-bearing field layer, while sites of higher performance contained extensive areas of food-bearing coniferous forest. Given such differences in availability, actual differences in patterns of habitat use

Table 4.7 Seasonal habitat use of New Forest roe. Figures show percentage of all animal observations in any season recorded in different habitats (from Sharma 1994).

	Winter	*Spring*	*Summer*	*Autumn*
Mature deciduous woodland	26	18	20	37
Young deciduous woodland	9	8	4	11
Mature coniferous woodland	36	48	36	33
Coniferous woodland (thicket/polestage)	13	4	9	7
Coniferous woodland (plantation/prethicket)	16	19	19	6
Rides/glades	<1	3	12	6

observed in the different sites were less than might have been expected: roe in poorer areas appeared to try to compensate for reduced availability of 'better' habitats by increasing the degree of positive selection for these. Despite this increased strength of selection, however, there remained significant differences in the pattern of habitat use observed in the different areas – with roe in poor-performance sites still observed to spend more time in food-depleted habitats than those of higher-density sites.

These differences in patterns of habitat use observed for roe in 'low'- and 'higher'-performance sites are equally reflected in differences in dietary quality. Roe at poor sites appeared to take significantly more pine needles and ericaceous browse throughout the year than do animals of higher-performance sites. Roe at better sites eat less grass and more bramble through summer and autumn than do animals at poorer sites, and as autumn progresses, make more use of ferns; they begin to increase their intake of heather far later in the year than roe at poorer sites and intake levels never rise as high (Sharma 1994). These differences between areas in the expressed pattern of habitat use – and in composition and quality of diet – differences which persist despite significant changes in the strength of selection observed for 'better' habitats, are considered by Sharma to be related in turn to differences in density and performance of roe deer in the different areas.

A number of authors have suggested recently that roe in other areas are able to compensate fully for differences in habitat availability, or differences even in the habitat types available, maintaining similar levels of performance and recruitment in different areas, despite gross differences in habitat use and diet. Thus, for example, recently both Calder (1995) and Wahlstrom (1995) have examined roe deer populations in forested areas by comparison to populations outside the forest block, resident within adjacent farmland areas. Both demonstrated marked differences in patterns of habitat use and diet of roe in the different environmental types – reflecting differences in the types of habitat available and their relative area within the range – but observed no differences in fecundity of individual females culled from woodland or farmland areas. However, it is clear that some response to differing resource levels is occurring: Wahlstrom, at least, noted that densities of roe were significantly lower (and the range size of individual females significantly higher) in the forested areas where resources were less rich. The animals appear to be able to maintain constant *per capita* reproductive rates by adjusting range sizes in areas of lower resource availability, such that the increased range still contains effectively the same absolute quantity of resources (see also Johnson and Putman, in preparation), distributing themselves across their environment in effect, in accordance with the ideal free distribution (Fretwell and Lucas 1970, Fretwell 1972).

In the New Forest too, densities of roe are lower where resources are less abundant; fewer animals 'choose to settle' in less preferred areas of poorer resource quality/availability and the carrying capacity of such areas is, in any case, reduced. But in these, strictly marginal, areas performance as well as density appears also to be affected: Sharma showed a clear correlation across different sites in the New Forest between the observed fawn/doe ratio and density of roe (Sharma 1994; Figure 4.2), indicating that where resource depletion has resulted in a reduced density of roe, reproductive performance of those animals is also reduced.

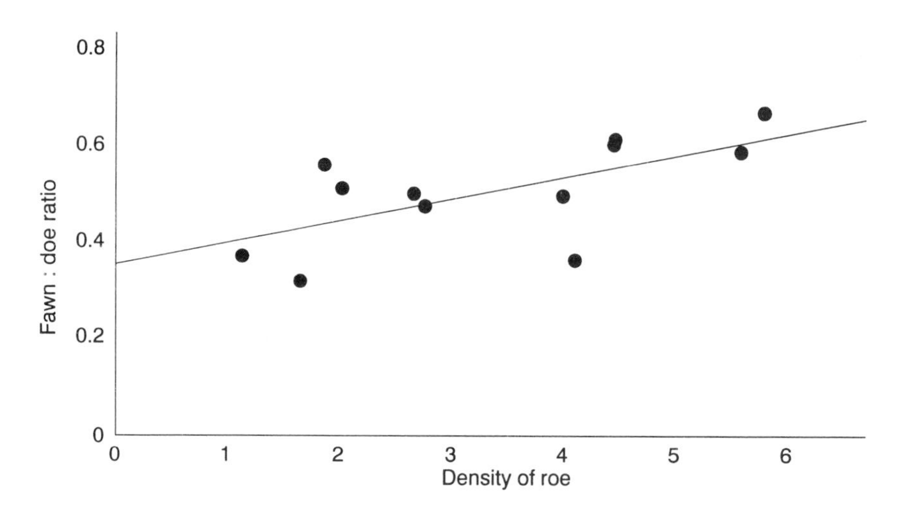

Figure 4.2 Recruitment rates in New Forest roe deer (as post-winter ratio of fawns to adult does) in relation to population density. Data (from Sharma 1994) consider variation between separate New Forest beats in mean fawn:doe ratio recorded over the period 1976–88 in relation to the mean density of roe in the same beats.

Such observations also go some way towards explaining the decline in numbers of New Forest roe over recent decades (p. 24). Those habitats which were present in far greater proportional area in sites containing low densities of roe of poor performance – and which thus featured more prominently in the pattern of habitat use expressed by those animals (despite the increased strength of selection for more preferred habitat types) – are the same habitats which may be seen to have increased in their overall area within the New Forest over recent years, at the expense of the available area of more productive ground. It seems entirely possible that the increase in relative area of these less-favoured habitats, associated with poor performance, may well lie behind the decline in populations of roe within the Forest as a whole.

Following the end of the Second World War, the late 1940s and early 1950s saw extensive replanting of the Forest woodlands for timber. Nearly all these new plantings were of conifer (while previously about three-quarters of the enclosed Forest comprised hardwood trees) and very large areas were afforested over the same brief period. Roe populations boomed; local keepers consider that the population of roe in the Forest in the 1950s was higher than it has been at any time subsequently. Most of these areas, which once supported such strong populations, have now reached the stage of canopy closure or beyond; and in the meantime, there has been little new planting.

Among coniferous plantations in general, younger plantings will support productive populations of the 'pioneer' roe. The wide spacing of the young trees allows development of grass and forbs in abundance – with other ground-layer species such as bramble or rose, known to be favoured by roe; the trees, too, are at a height where roe can, if they wish, browse on the growing shoots. As the plantation matures, the canopy closes and cuts out the light reaching the forest floor; the ground flora is eliminated and the trees themselves are too tall to offer much forage. Roe are still found in such plantations, but use them almost exclusively for cover, foraging out from such shelter belts into younger woodland blocks or open vegetation (rides or clearings) at dawn and dusk.

As a result, such plantations can support good populations of roe only if they are associated with productive open ground nearby, or other, younger blocks of woodland. In most commercial coniferous forests, trees are planted in blocks and on a regular rotation, so that at any one time, a single forest will contain blocks of trees at all stages, from areas newly planted to mature stands of fully grown timber. But as we have noted earlier: the Forest has had a rather curious management history, and large areas of previously prime roe habitat have now matured to a stage where they offer only limited resources, while relatively little new planting has been undertaken in recent years to provide alternative foraging areas elsewhere.

Sharma (1994) has investigated this more formally, relating changes over time in the population size of roe from 1975 to 1985, to changes in a number of environmental parameters over the same period (changes in vegetational structure and composition, climatic conditions, density of other deer species and Common stock). Preliminary results were reported by Putman and Sharma (1987).

Population estimates used in these analyses were those population figures returned by Forestry Commission keepers in the annual spring census (see Chapter 2 and Table 2.1). These figures are clearly not an accurate measure of absolute number, nor are they assumed to be. However, as a relative **index** of population change from year to year, these census figures are reliable – since they have over that 10-year period relied on the same method and the same observers within each Forest 'beat'.

Surveys of the Forest timber stocks are also undertaken by the Forestry Commission on a regular basis. The Commission periodically carries out detailed inventories of all enclosed woodland blocks, recording for each the tree species present, their date of planting and their estimated growth rate. This information is used to provide stock maps and timber production forecasts, and is continually updated as trees are felled or thinned and areas are replanted. Data collected provide details only of timber species planted and height, but estimates of area are accurate to 0.5 ha. From these records it is possible to extract information on several vegetational parameters that might affect the Forest deer. (Stock surveys are only undertaken every 5–7 years and thus data are not directly available for the intervening years. However, an estimate of forest structure in any year other than those for which specific records are available may be obtained by extrapolation; a computer model developed by the Forestry Commission for forward projection of forest structure (Ratcliffe *et al.* 1986) was used to 'recreate' the structure of each Forest beat in years for which no direct survey data were available (for detailed methodology, see Putman and Sharma 1987).)

Roe deer population number at any time was found to be independent of its own previous size/density – either in the previous year or average density over the immediately preceding 3-year period; roe numbers were equally found not to be strongly correlated with densities of other sympatric ungulate species (Chapter 6). However, observed changes in the roe deer population were very strongly correlated with changes in habitat character and also with overwinter climate. Specifically, censused roe numbers were positively correlated with the available area of broad-leaved and coniferous woodland blocks of between 2 and 5 m in height ('prethicket stages' with dense understorey, before closure of the canopy shades out the shrub and field layers), declining as woodlands matured. Roe numbers showed significant negative correlation with total area of conifers between 5 and 15 m and with the area of conifer and broad-leaved trees greater than 20 m in height – the same variables associated with poor performance in present-day populations.

Roe numbers censused in April in any year were found also, independently, to show significant positive correlation ($P<0.05$) with mean March temperatures of the previous year and total precipitation in February of the current year; April population estimates were negatively correlated with mean daily temperature over the immediately preceding winter (November to February), and mean temperature and rainfall over the previous summer (specifically May/June; $P<0.05$). However, in multivariate analyses undertaken to explore the relative importance of the various environmental factors found to account for variation in censused roe populations, habitat variables alone (and notably the available area of prethicket conifer, before canopy closure) accounted for 93% of the recorded variation in estimated roe populations, with climatic factors contributing only a further 4% to explained variation (Sharma 1994, Putman *et al.* 1996).

5

The domestic stock of the New Forest

5.1 THE HISTORY OF COMMON PASTURAGE

Since its afforestation in the eleventh century, and very probably even before that time, domestic animals have been pastured on the Open Forest grazings alongside the Forest deer.

Although records before the initial Registers of Claims for Common Rights in 1635 and 1670 offer only very fragmentary evidence about the history of commoning in the years leading up to the seventeenth century, it is clear from what evidence is available (presentments at Forest Courts in respect of abuses of Common Rights, or petitions from Commoners themselves alleging unreasonable infringement of these rights) that the area supported a vigorous pastoral economy. Sheep, cattle and ponies were grazed upon the Forest from the thirteenth to the sixteenth centuries. Some of the biggest graziers were the large landowners – the big religious houses of the surrounding area – but at the same time it is clear that all Rights of Common were also extremely important to occupiers of smaller holdings: cottagers who could make a subsistence cottage economy considerably more comfortable by exercising their rights to fuel, animal bedding and fodder for their few livestock. With the dissolution of the monasteries in the middle of the seventeenth century and the accompanying changes in land ownership more generally, patterns of commoning began to change. Fewer sheep were depastured (sheep flocks in the past having been for the most part associated with the big religious houses); pasturage of livestock was primarily of cattle and ponies, with continued seasonal pannage of pigs. Further, with the disappearance of the large landholdings of the monasteries, there was a shift in the entire social economy of commoning. In the Registers of Claims compiled in 1635 and 1670, most claims were in respect of cottagers with only a few acres; even the holdings of tenant or yeoman farmers with land adjoining the Forest or within the Forest perambulation (and thus entitled to exercise their Common Rights) were usually of between 15 and 30 acres (6–12 ha). Increasingly too, Commoners were exercising their rights not just in order to support a cottage economy, but to support agriculture. Tubbs (1968) in his excellent *The New Forest: An Ecological History*, notes that the majority of holdings were actually too small in themselves to be self-supporting, and we may deduce that grazing rights in particular were being used to support an economy of small livestock farmers, entirely dependent of the use of Common lands, rather than simply to support the few beasts of a subsistence cottage economy.

The picture remained essentially unchanged through the eighteenth and nineteenth centuries. The majority of claims entered for Rights of Common were typically for freeholdings of less than 80 acres (30 ha). Of 1200 holdings registered in 1858, 629 were for properties of less than 30 acres and 400 of these were for cottages with between 1 and 4 acres (0.4–1.6 ha). In 1944, Kenchington quoted the total number of holdings with Common Rights upon the Forest as 1995; of these 731 were below 5 acres (2 ha) and only 326 were above 50 acres. Two-thirds of the holdings represented the main sources of livelihood for the occupiers. Clearly the shift in use of Common Rights away from merely assisting a cottage economy towards supporting grazing agriculture on a substantial scale had continued still further.

Numbers of cattle and ponies depastured on the Forest today are as high as those of the peak years of commoning in the 1880s (Figure 5.1), but the actual number of Commoners exercising grazing rights is considerably smaller. While 1200 individual holdings were registered in 1858, today Rights of Common are exercised only by a few hundred. The picture is still one of many small farmers, turning out small numbers of stock, rather than a Common dominated by a few large landowners turning out huge herds, but now only 10% of those holdings are worked full-time and for half the Commoners included in a recent survey (Countryside Commission 1984) their 'agricultural enterprise' contributed less than 10% of their total income. Cattle and ponies predominate among the stock turned out on the Forest grazings. A few sheep are turned out in local areas and in other areas a few donkeys. Pigs are still put out to forage for acorns in the Forest woodlands in autumn, but the major grazers of Open Forest remain the cattle and ponies.

Although free-ranging over the unenclosed Forest, the cattle and ponies of the New Forest are all privately owned, and all more or less closely managed. Few animals are simply turned out on to the Forest to fend for themselves; the majority of Commoners 'look over' their stock on a regular basis, worming animals as required, removing from the open grazings animals clearly in poor condition. The Common grazings are also 'policed' by the four Agisters – statutory employees of the Court

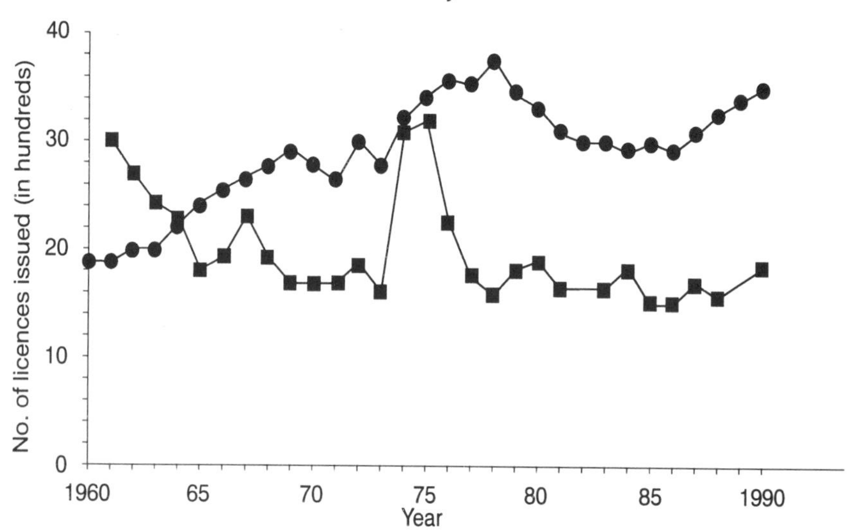

Figure 5.1 Recent changes in abundance of cattle (■) and ponies (●) licensed to graze on the Open Forest.

of Verderers which controls the Forest Commons, whose main duty is to ride regularly throughout the area checking on the condition of the Common stock, reporting to the Verderers and individual owners on the condition of the grazings and the animals – and with the power themselves to order off the Forest animals considered to be in poor condition.

The cattle are perhaps more closely managed. Few are entirely free-ranging within the Forest; most are turned out in the immediate vicinity of the owner's own holding and are regularly handled. Many indeed are turned out only during daylight hours and return to the holding at night. Relatively few are left on the Forest grazings year round; those which are left out over winter are offered supplementary forage in the form of hay or straw in the vicinity of the holding to which they belong. The ponies range more widely over the Forest during spring and summer. However, in the autumn of each year all ponies, too, are rounded up in a series of 'drifts' in different areas of the Forest; young of the year are branded, animals are sorted and those destined for sale – foals of the year or older animals – are removed. Animals due to be returned to the Forest are tail-clipped with the owner's mark, wormed and released. While many more ponies than cattle remain on the Forest grazings through the winter, a number of older animals are removed during the autumn drifts, overwintered on the owner's holding and returned to the Common grazing the following spring. Those left upon the Forest depend largely on natural forages, although some receive some supplementary hay or straw – and others learn to frequent areas where such supplements are put out for cattle.

For much of the year then these domestic animals exploit the forage resources of the Forest alongside the Forest deer; although many cattle and some ponies are removed from the open Commons during late autumn, a number of cattle and a substantial population of ponies do remain on the Forest over winter, dependent largely on the limited natural foraging available.

5.2 SOCIAL ORGANIZATION AND BEHAVIOUR

Studies of the social organization of most populations of wild or semi-wild horses (e.g. Feist and McCullough 1976, Berger 1977, 1986, Wells and von Goldschmidt-Rothschild 1979, Rutberg 1990, Duncan 1992) have shown that the normal social 'unit' is a stallion-maintained harem of mares and subadults: a harem which is maintained throughout the year. Stallions are aggressively territorial and defend not only their group of mares, but also an exclusive home range.

The ponies of the New Forest, while equivalent in many respects to these truly feral populations, do show certain characteristic differences as a consequence of their more intensive management. Groupings are often artificial, resulting as fixed associations of animals belonging to a single holding which were put out on to the Forest together and retain their close association. In addition, many owners, as we have noted, do not leave their animals out all year round, but take them off the Forest in the winter, only returning them to free-range the following spring; such a practice reinforces each year the social bondings of animals of a given holding. Not all owners do take their stock in over the winter period, and many mares and foals do remain on the Forest throughout the year. None the less, 'natural' groupings are difficult to maintain: while many owners may be prepared to leave their mares and foals on the Forest all winter, few indeed allow the valuable stallions to overwinter. With a few exceptions, stallions are run on the Forest only from April to September. As a result, natural harem groups are rare. Females may form permanent associations, which are maintained throughout the year; a stallion may join up with such a group over the summer months, but he plays little role in social cohesion; the groups are not true harems and the social order is essentially matriarchal.

The social structure of New Forest ponies is thus somewhat atypical, with the basic social unit most usually observed being a mare–foal assembly (Tubbs 1968, Tyler 1972). The fundamental unit within such a grouping may be recognized as an adult mare with her offspring of current and possible also previous years (Tyler 1972). Larger groups may be formed as an association of these basic units – as older offspring maintain their association with a matriarchal group, or as 'artificial' associations are forged between animals from the same holding, turned out on to the

Forest together, but the size, cohesiveness and persistence of these social groups is extremely variable. Around this basic structure other, more temporary associations are also observed. Larger aggregations may occur on favoured feeding grounds, associating casually, simply because they happen to be in the same place at the same time. Further, groups of animals using the same home range, frequently seem to use the range in the same way on a diurnal basis. This may give rise to what appear to be larger groupings of animals, but such aggregations are not true social groups and do not persist for long. Stallions, when on the Forest, may also form the focus of larger aggregations of mares, ranging in size up to 10–15 adults and juveniles. These harems too are simply composed of assemblages of mare–foal units, and these matriarchal groups join or leave the harem as a cohesive unit (Tyler 1972, Pollock 1980, Putman 1986b).

The majority of pony groups within the Forest may be ascribed to a population occupying the same or strongly overlapping home ranges. Such ranges are always centred upon a reseeded or improved grassland, streamside lawn or other area of rich grazing. Since such areas are relatively few within the Forest as a whole, the populations are fairly discrete and few individuals or groups move regularly between populations (see also Feist and McCullough 1976, Rutberg 1990). The size of the home range varies markedly over the Forest and seems primarily to be determined by the distribution in space of four essential component resources: a suitable grazing area, a water supply, shelter (usually provided by woodland or gorse) and a 'shade' (Tyler 1972) – a traditional resting area, where groups of ponies aggregate in summer for protection from sun and biting flies.

Depending on the dispersion of these resources in space, ranges vary hugely in size. In the north of the Forest, where vegetation communities occur in relatively large, homogeneous blocks (coarse-grained), these components of the home range may be widely separated and consequently large ranges may be found, often exceeding 1000 ha (Putman *et al.* 1981); further south, a more heterogeneous vegetation cover results in smaller home ranges, down to 100 ha or less.

Within this home range we can appreciate a general pattern of range use. Over the summer period, each pony generally spends most of its time feeding on the grazing area(s) included within its range. During the course of the day, there is at least one move off the lawn to drink, and at this time the animals usually make some use of the lusher forage available in the wetter vegetation types. At dusk, there is a general move away from the more exposed grazing areas, to vegetation types offering more cover, and the ponies usually spend the hours of darkness in, or in the shelter of, areas of woodland or gorse. Those groups of animals left on the Forest over the winter make somewhat less use of the exposed grassland communities, although these are still grazed extensively – despite the fact that there can be little useful forage left. However, they make greater use over winter of those areas of their range offering shelter from the more severe weather, and emerge from the woodlands or gorse scrub for shorter periods during the hours of daylight. Range size is often reduced during the winter months, since water is more readily available and there is no need to 'shade' (Putman *et al.* 1981, Putman 1986b).

Cattle are now run on the Forest mainly for beef production and are mostly Friesian/Hereford or more mixed crossbreeds. Herds are composed predominantly of 1- and 2-year-old heifers or mature beef cows and calves, these groups being commoned separately by their owners. In contrast to the ponies, the basic social unit of cattle in the New Forest is the herd, although the cohesiveness of this unit varies seasonally. The dispersion of these herds across the Forest reflects to some extent the location of the farmsteads, since many owners release their animals in the immediate vicinity of the holding (and, as noted, some allow them to return to enclosed fields at night). As in the case of pony 'populations' therefore, cattle herds belonging to different owners are dispersed over the Forest in relatively discrete units, with each unit having a home range centred on one or perhaps more primary grazing areas.

Over the summer months, the herds fragment into a number of smaller groups which range independently but encounter each other frequently on grazing sites and commonly aggregate together in 'camps' overnight. These small subunits are not cohesive and individuals regularly move from one to another. The pattern of habitat use over this period is similar to that of the ponies, with cattle spending most of the daylight hours on the primary grazing areas, although making more frequent visits to water supplies. Less use is made of wetland vegetation and the cattle appear to avoid particularly boggy areas. By night there is again a general move to those vegetation types offering cover, particularly deciduous woodland, although groups will often spend the night out on dry heathland when weather conditions are mild.

During this summer period, cattle range far more widely over the Open Forest than do the ponies, often occupying home ranges with more than one primary grazing site; these areas may be used in turn over the course of several days and daily movements of 4 or 5 km are not uncommon. Movement is particularly common at dusk and dawn, with groups often moving 2 km or so in deliberate, purposeful 'marches' from feeding site to overnight camp. (These distances, although large in comparison to most pony-group movements, are less than have been recorded in similar studies of cattle on native range. Schmidt (1969), for example, followed shorthorn cows over 9.2 km/day on open range; Shepperd (1921) recorded daily movements of 8.9 km within a 260-ha paddock. The difference may be explained by the relative heterogeneity of the habitat mosaic and availability of water in the Forest environment (Putman 1986b).)

Over the autumn, the majority of cattle are removed from the Forest; those few herds that do remain are generally fed to a considerable extent with hay and straw feed. This practice of supplementation radically alters the entire activity pattern and ranging behaviour of the animals. Subunits of the herd are drawn together at the feeding site and herd cohesiveness is markedly higher over the winter months than during summer. Feeding usually takes place at a fixed site, 1 or 2 hours after dawn. The animals congregate at the site over this period and stand around ruminating or resting; little foraging occurs.

With the arrival of the fodder, the cattle feed steadily till all is consumed, and then lie ruminating for the rest of the morning and early afternoon. Some sporadic feeding on natural vegetation may occur, but intake is negligible. About 1 hour before dusk there is a general departure from the feeding site and the herd makes its way to the selected overnight area. While the route taken may vary, the site selected to spend the night is almost always the same, and this often involves considerably greater distances of movement than those recorded during the summer months; round trips of 11.5 or 12 km have been recorded regularly (Putman *et al.* 1981; Putman 1986b). While *en route* the animals feed extensively, mostly on *Calluna* on wet and dry heathland communities; feeding may continue within the overnighting area over the first hour or two of darkness, particularly if a grazing area is available, before the animals bed down for the night. At first light the herd begins the return march, on which little foraging takes place, and arrive at the feeding site in good time for next day's fodder provision.

5.3 PATTERNS OF HABITAT USE

Such descriptions as those just offered of range use are of course crude, over general and oversimplistic. Detailed studies of the pattern of use of different vegetation types by ponies and cattle were undertaken as part of a programme of research sponsored by the (then) Nature Conservancy Council between 1977 and 1981 to investigate any possible deleterious impact of increasing grazing pressure from Common stock on the Forest ecology. Results of these more detailed studies have been published by Putman *et al.* (1981, 1984) and Pratt *et al.* (1986), and confirm that for the **ponies** the various different grasslands of the Forest (and in particular the different improved grasslands and streamside lawns) are tremendously important communities throughout the year (Table 5.1). In direct observations, the proportion of animals recorded on improved grasslands and streamside areas (as a proportion of the total numbers observed on all communities at any one time) remains throughout the year in excess of 40–50%. Increasing stock numbers over spring and summer, coupled with the effects of drought, result in reduced availability of grass over the summer months. Declining use of these grasslands is compensated for by increased use of the Forest bogs and wet heath; in autumn and winter, increased use is made, both for shelter and in foraging, of deciduous woodland and gorse brake.

Table 5.1 Habitat use of New Forest ponies (1978–79) based on data of Pratt *et al* (1986). Figures show percentage of all animal observations in any month recorded in different habitats.

	Jan.	*Feb.*	*Mar.*	*Apr.*	*May*	*June*	*July*	*Aug.*	*Sept.*	*Oct.*	*Nov.*	*Dec.*
Deciduous woodland	20	12	12	10	6	6	6	9	13	13	7	6
Coniferous woodland (mature)	0	0	0	0	1	3	2	1	2	0	0	2
Rides and glades	0	4	0	2	0	2	3	2	6	2	4	5
Gorse brake	12	16	2	1	0	1	0	0	0	3	3	11
Dry heath	4	10	4	3	14	11	7	7	12	11	18	11
Wet heath	15	9	7	7	14	14	13	13	13	7	5	8
Bog	2	5	7	9	9	12	8	10	12	4	4	4
Acid grassland	3	9	1	10	3	4	6	4	3	2	3	2
Streamside lawn	5	6	12	9	19	12	13	9	9	11	13	3
Improved natural grasslands	33	15	27	22	14	11	18	27	18	34	23	27
Reseeded areas	2	5	17	21	14	17	19	12	8	10	11	16
Road verges	4	8	11	6	6	6	4	5	5	2	9	6

Cattle were observed in all community types used by ponies; like the ponies they are preferential grazers, and patterns of habitat use recorded reveal for them, too, the importance of the various grassland communities (Table 5.2). The various improved grasslands, together with stream margins, sustained in excess of 40% of all cattle observations throughout the year, with over 70% use during the summer months (July–September). Heathland communities were also used extensively throughout the year, except in midwinter. Deciduous woodland showed peak use in spring and autumn; regenerating heathlands and bogs showed peak use from March to May. Overall, the patterns of habitat use throughout the year showed relatively little variation, and one of the most striking differences between cattle and ponies noted by Putman *et al.* (1981) and Pratt *et al.* (1986) was this comparative lack of seasonal variation in habitat occupancy by cattle.

Table 5.2 Habitat use of New Forest cattle (1978–79) based on data of Pratt *et al.* (1986); figures show percentage of all animal observations in any month recorded in different habitats.

	Jan.	*Feb.*	*Mar.*	*Apr.*	*May*	*June*	*July*	*Aug.*	*Sept.*	*Oct.*	*Nov.*	*Dec.*
Deciduous woodland	0	39	10	10	16	0	9	2	8	24	2	0
Coniferous woodland (mature)	1	0	1	0	0	0	0	1	6	0	0	0
Rides and glades	0	18	0	1	0	1	1	0	0	1	5	0
Gorse brake	13	0	0	1	2	1	1	0	2	0	0	0
Dry heath	0	0	12	5	3	24	2	9	3	3	17	11
Wet heath	0	5	5	16	9	1	6	2	7	10	7	6
Bog	0	1	0	4	5	0	2	0	2	0	0	11
Acid grassland	9	4	0	2	2	1	0	4	0	1	3	4
Streamside lawn	24	0	8	6	16	0	5	9	5	0	2	11
Improved natural grasslands	2	0	40	17	19	0	37	16	21	16	34	0
Reseeded areas	51	29	19	18	24	73	31	55	43	42	30	52
Road verges	0	3	4	1	3	0	11	1	3	4	0	4

However, both species do display a pronounced diurnal shift in habitat use, with daytime habitats selected primarily for feeding and night-time selection for communities offering some degree of cover. Diurnal movements between vegetation types by ponies are very much more marked over the summer months than during winter. Throughout the year, the ponies show preferential night-time use of cover communities such as gorse brakes or areas of deciduous woodland. In winter, these 'shelter' communities are also used more extensively during the daylight than they are during summer. Thus over the winter period, the diurnal switch between daytime and night-time communities is less pronounced since the ponies spend much of their time in more sheltered areas making only occasional forays out on to the more open vegetational communities. In summer, however, most of the daylight hours are spent in open habitats (grasslands, heathlands, bogs) and diurnal movements into shelter habitats at night are much more obvious. Differences between daytime and night-time patterns of habitat use are even more marked amongst cattle – and persist throughout the year. In all seasons, the cattle spend most of the daylight hours on open vegetation communities, making a deliberate movement into other, shelter communities at dusk and back to open grazing areas at dawn (Putman 1986b).

Habitat preferences and individual patterns of habitat use differ between individual animals. What we have described here is an overview of the pattern of use of vegetation by a whole population; within this there is considerable individual variation (Howard 1979, Gill

1988). In part, such differences are imposed upon the animals – as for Sharma's roe deer (Chapter 4) – because of differences in the relative availability of the various vegetational components in different geographical regions of the Forest. Thus pony populations in areas without extensive gorse brake will make more use of woodland for winter shelter, and ponies in areas with no improved grasslands make more use of streamsides or roadside verges, natural acid grasslands or heath.

But this is not a complete explanation – and even among individuals who share the same home range, there are subtle differences in the extent of use of different habitats; such individual differences clearly reflect fundamental differences in individual habitat preferences and may have considerable implications in relation to differences between animals in the ability to maintain body condition (Gill 1988, and below, pp. 84–5). In practice, however, such differences in habitat use are rather slight. Studying patterns of habitat use by Forest ponies in four different sites chosen specifically because they did differ markedly in vegetation and standing crop, Gill (1988) found that habitat use by the ponies in the four areas was actually very similar. Patterns of habitat use revealed in her analyses (Table 5.3) are also extremely similar to those reported earlier by Pratt *et al.* (1986); such differences as appear are consistent with differences in availability of the various vegetational communities in different study areas used – and the fact that in contrast to the earlier studies of Pratt *et al.* (1986), who recorded habitat use over the full 24-hour period,

Table 5.3 Habitat use of New Forest ponies (1982–84) based on data of Gill (1988). Figures show percentage of all animal observations in any month recorded in different habitats. (Note: Gill made no distinction between areas of improved grassland established by bracken clearance and liming and those established by reseeding.)

	Jan.	*Feb.*	*Mar.*	*Apr.*	*May*	*June*	*July*	*Aug.*	*Sept.*	*Oct.*	*Nov.*	*Dec.*
Deciduous woodland	3	5	5	1	1	0	0	0	0	0	<1	<1
Coniferous woodland (mature)	<1	<1	0	0	0	0	0	0	0	0	0	0
Rides & glades	3	6	6	3	3	2	3	1	2	2	3	2
Gorse brake	36	29	19	11	2	1	0	0	4	3	4	14
Dry heath	1	1	1	1	<1	0	<1	<1	1	1	1	1
Wet heath	6	6	6	3	4	8	3	13	12	11	6	9
Bog	3	2	5	5	3	2	1	4	8	2	2	1
Acid grassland	11	11	11	13	19	15	15	21	18	18	22	19
Streamside lawn	3	4	3	4	5	3	4	7	4	3	4	4
Improved grasslands	24	24	35	47	51	53	52	36	41	46	43	33
Road verges	8	8	7	11	11	12	12	13	11	14	13	15

Gill's analyses were based on observations during the hours of daylight only, and thus likely to under-represent use of cover habitats such as coniferous and deciduous woodland.

5.4 DIET

Despite minor differences in the patterns of habitat use shown by cattle and ponies, the overriding impression is one of striking similarity. Both species are preferential grazers, feeding primarily on the various improved grasslands of the Forest and, overnight, seeking cover in woodlands or gorse brake. Analyses of diet reinforce the same impression of similarity; Tables 5.4 and 5.5 summarize dietary profiles in terms of percentage species composition of identifiable plant fragments in monthly faecal samples (from Putman *et al.* 1987).

A glance at Table 5.4 reveals the essential simplicity of the diet of the Forest cattle; in effect they fed throughout the year on a restricted range of grasses and made up the balance with heathers. Consistent with their seasonally unchanging use of habitat, the diet also remained remarkably similar throughout the year: varying only in minor changes in relative proportion of these two main dietary components. Throughout the summer months (May–August), approximately 80% of the diet was found to be made up of grasses (primarily *Agrostis capillaris, A. canina, A. curtisii* and *Festuca rubra*); heathers (both *Calluna* and *Erica*) together contributed a further 14% overall. No other items contributed more than 1–2% of the diet at this time. Through autumn a small reduction was seen in the percentage of grasses taken, with corresponding increase in heather intake.

Cattle in both our study areas were fed hay or straw from November to March (as is indeed usual for all cattle left on the Forest overwinter). Hay and 'fresh' grass are indistinguishable in our analyses: together they continued to make up some 70–75% of the diet throughout the winter months, with heather contributing overwinter a further 20–25% of the total (Putman *et al.* 1987). In effect, without the distinction between hay feeding over winter and grass feeding in the summer months, no significant change at all could be detected in diet through the year, merely as noted, a slight change in relative emphasis of the grass/hay and heather components.

By contrast pony diet showed marked seasonal change, with characteristic winter and summer forages and marked transition periods in spring and autumn (Table 5.5). Through the summer, the animals fed extensively on grasses, which comprised between 80 and 90% of the diet at this time. By contrast to cattle, the ponies of Putman *et al.*'s studies fed extensively on *Molinia* over this period: *Molinia* makes up nearly 20% of the total diet recorded, 22% of grass intake at this time.

Table 5.4 Diet of cattle in the New Forest (1978–79) from faecal analyses of Putman *et al.* (1987); figures show percentage contribution to the diet of different forages consumed.

	Jan.	*Feb.*	*Mar.*	*Apr.*	*May*	*June*	*July*	*Aug.*	*Sept.*	*Oct.*	*Nov.*	*Dec.*
Molinia	0	0	0	0	0	0	1	0	1	1	1	1
Other grasses	75	67	71	80	80	83	81	70	65	69	69	65
Forbs	0	0	0	t	1	t	0	0	1	1	t	1
Sedges/rushes	0	2	t	t	0	0	1	1	1	1	1	1
Conifer browse	0	0	0	0	0	0	0	0	0	0	0	0
Holly	0	0	0	0	0	0	0	0	0	0	0	0
Other broad-leaves	1	1	1	1	0	0	0	1	2	1	2	1
Heather	21	27	24	9	14	12	13	18	23	19	21	22
Bramble/rose	0	0	0	0	0	0	0	0	0	0	0	0
Gorse	0	0	0	0	0	0	0	0	0	0	0	0
Mosses	1	2	3	7	3	3	3	5	6	5	3	8
Ferns	2	1	1	1	2	1	1	3	1	2	2	1
Others	0	0	0	2	1	1	0	2	0	1	1	0

t = Trace.

Table 5.5 Diet of ponies in the New Forest (1978–79) from faecal analyses of Putman *et al.* (1987); figures show percentage contribution to the diet of different forages consumed.

	Jan.	*Feb.*	*Mar.*	*Apr.*	*May*	*June*	*July*	*Aug.*	*Sept.*	*Oct.*	*Nov.*	*Dec.*
Molinia	2	1	t	t	17	22	24	17	7	3	2	2
Other grasses	48	36	43	65	73	68	68	70	76	76	67	49
Forbs	0	0	0	0	0	t	t	t	0	0	0	0
Sedges/rushes	2	2	2	2	3	3	3	2	2	2	2	2
Conifer browse	0	0	0	0	0	0	0	0	0	0	0	0
Holly	19	26	25	13	0	0	0	0	0	3	11	13
Other broad-leaves	0	0	0	0	2	1	1	t	1	0	0	0
Heather	7	7	5	3	1	t	1	1	2	3	6	10
Bramble/rose	0	0	0	0	0	0	0	0	0	0	0	0
Gorse	12	13	10	1	0	0	0	0	t	1	3	9
Mosses	9	13	14	15	3	4	1	2	2	5	7	12
Ferns	1	2	1	t	0	2	2	7	10	7	2	3
Others	0	0	0	1	1	0	0	1	0	0	0	0

t = Trace.

During September and October, intake of *Molinia* (a markedly seasonal, deciduous grass) declined to only 3%; however, total grass percentage remained relatively constant at around 80%, with a greatly increased intake of *Agrostis curtisii* balancing the decline in consumption of *Molinia*.

Considerable use was also made over this period of bracken (*Pteridium aquilinum*). Overall the diet can be shown to differ significantly from that of the summer, although this is a progressive change (Putman *et al.* 1987).

As autumn changes to winter, the percentage of grass in the diet is found to decline, to 50% of total intake. Although some ponies do use the hay or straw put out for cattle over the winter, it does not form so significant a part of their diet as it does for the cattle themselves. Instead a progressive increase was recorded, from October right through to February/March, in the amount of gorse (*Ulex europaeus*) and tree leaves (mostly holly: *Ilex aquifolium*) which were taken. The proportion of moss fragments found in the faeces also increased over this period and heather intake also increased in winter (mean 6.0% November–February). These changes are progressive right through the winter period; at the end of winter a further transition is evident (Table 5.5) in a gradual change through spring back towards the diet characteristic of the summer months. This spring diet, showing a gradual increase in grass intake, but still significant levels of gorse, moss and tree leaves, differs significantly from both true winter and summer diets.

The data of Table 5.5 summarize diets recorded for ponies in two different areas of the Forest, between January and December 1979 (Putman *et al.* 1987). An independent analysis of the diet of the Forest's ponies is presented by Gill (1988) based on analysis of faecal materials collected between May 1984 and April 1985 in four different sites within the Forest. Gill's sites were selected specifically to include areas varying markedly in habitat composition and vegetational standing crop; however, her areas included both sites sampled earlier by Putman *et al.* (1987). Dietary composition recorded by Gill substantially supports that described in the earlier study, with ponies found to feed predominantly on grasses in the warmer months of the year, increasing intake of browse during winter and early spring as availability of grass declines (Table 5.6). However, there are some significant differences in the dietary profiles reported by Gill, with lower recorded use made of *Molinia* in the summer months and greatly increased frequency of moss fragments in the dung at all times of year. In addition, Gill's results, derived from a wider range of study sites across the Forest, suggest that within the Forest as a whole, ponies perhaps make less use of holly and greater use of gorse as winter browse than suggested by the earlier, more geographically restricted studies of Putman *et al.* (1987).

Gill's detailed analyses reveal that there were significant differences in diets recorded in any season for the pony populations of different regions of the Forest: differences which reflect in large part differences in availability of different forage types (Gill 1988). While the figures of Table 5.6 thus provide a useful overview of the overall diet of ponies in the

Forest as a whole, the recognition of substantial regional variation emphasizes that there is no such thing as an average pony. Any interaction between the different herbivores of the Forest – particularly between cattle and ponies, will of course occur within particular areas or subpopulations of the Forest. Where analysis of Gill's data is restricted to the two areas used by Putman *et al.* in their simultaneous analyses of the diets of both ponies and cattle, the dietary profiles revealed are more similar. Even in these two 'common' study sites, however, Gill (1988) records a greater use of forbs, a greater intake of mosses, particularly during the winter, and significantly lower use of *Molinia* during the summer than that apparent in the earlier studies; suggested intake rates of gorse as winter browse are also considerably higher in Gill's analyses.

Table 5.6 Diet of ponies in the New Forest (1982–84) from analyses of Gill (1988); figures show percentage contribution to the diet of different forages consumed.

	Jan.	*Feb.*	*Mar.*	*Apr.*	*May*	*June*	*July*	*Aug.*	*Sept.*	*Oct.*	*Nov.*	*Dec.*
Molinia	0	0	3	1	0	t	t	1	t	0	0	t
Other grasses	26	23	29	38	55	72	62	60	50	50	55	39
Forbs	14	13	12	10	4	3	5	6	4	4	5	10
Sedges/rushes	t	t	1	3	3	3	4	3	1	2	1	t
Conifer browse	0	0	0	0	0	0	0	0	0	0	0	0
Holly	5	7	7	6	2	t	t	0	2	1	1	5
Other broad-leaves	0	0	0	t	1	t	t	1	1	1	0	0
Heather	12	10	5	4	4	1	1	2	5	3	3	7
Bramble/rose	0	0	0	0	0	0	0	0	0	0	0	0
Gorse	29	30	21	9	1	t	2	4	5	6	14	28
Mosses	15	17	22	31	30	18	14	11	13	20	20	10
Ferns	t	t	0	0	1	2	11	11	17	7	t	t
Others	0	0	0	0	1	1	1	1	2	6	1	1

t = Trace.

Diets reported here for New Forest ponies and cattle are very similar to those reported elsewhere for cattle and horses in a similar range (Putman 1991). Thus species composition of the diet of Exmoor ponies, and seasonal change in dietary composition closely mirror those reported here for the New Forest (Gates 1980, 1982). Exmoor ponies, like their New Forest counterparts, graze extensively throughout the year, showing a strong preference for *Agrostis* and *Festuca* swards but feeding to a significant extent upon *Molinia* during summer and autumn. When the availability of such grazing declines in the winter, the ponies of Gates' study compensated again with increased intake of gorse and heather. Such a pattern of forage use is strikingly similar to that observed for New Forest ponies, suggesting very similar strategies and responses in the two different populations.

Other studies of feral horse populations also show a dietary pattern essentially similar to that described for New Forest animals, with grasses dominating the diet throughout the year and other foodstuffs taken only when these preferred forages are less abundant. For example, the diet of horses in the Camargue during the growing season (spring and summer) consisted entirely of grasses, together with sedges and rushes from the adjacent wetland areas; only in autumn and winter, when grasses are less available, are less preferred species such as perennial herbs, halophytic forbs and coarser grasses consumed (Duncan 1983; Mayes and Duncan 1986). In western Alberta, up to 90% of the diet of feral horses during spring and summer is composed of grasses, with some sedges; very few forbs are taken at any time of year, and only in late winter are browse species such as pine and spruce taken in any quantity (Salter and Hudson 1978, 1979). Similar trends are reported for other North American horse populations (Feist and McCullough 1976, Hansen 1976, Berger 1986).

5.5 INDIVIDUAL VARIATION IN PATTERNS OF RESOURCE USE AND CYCLES IN BODY CONDITION

Gill (1988) reports clear differences in diets selected by different individual ponies; she also notes considerable individual variation in habitat preference and habitat use. In part, as we have noted, these differences reflect differences in relative availability of different habitats and forage types in different areas within the Forest, but in practice differences in patterns of resource use between individuals of different geographical area are less than might be expected on this basis (Gill 1988), and certainly no greater than individual differences observed between animals occurring in the same area. Both Howard (1979) and Gill (1988) record substantial variation in patterns of resource use even between individuals sharing the same home range. This variation is significant (statistically and biologically) – in that some of these individual differences in patterns of habitat use or diet are actually reflected in consistent differences between individuals in performance and overall body condition.

In common with many other feral temperate ungulates, New Forest ponies display a pronounced seasonal cycle of body condition: building up condition and fat reserves through summer and autumn, and relying on those reserves to make up any short-fall in energy balance through winter and early spring. Thus Pollock (1980) remarked a general increase in condition of New Forest mares through late spring, summer and autumn, with a general deterioration in condition apparent from mid-winter until spring (also the time at which mortality was highest).

Such cycles of condition have also been noted in other feral horse populations by Berger (1986) and Mayes and Duncan (1986) – and might be taken as evidence to suggest resource limitation, with declines in condition overwinter reflecting a need to draw on body reserves to make good a short-fall in energy intake available from available forage. However, similar cycles of condition in various species of ruminants (e.g. red deer, *Cervus elaphus* (Mitchell *et al.* 1976, Kay 1979, Kay and Staines 1981); white-tailed deer, *Odocoileus virginianus* (McEwen *et al.* 1957, Short *et al.* 1969, Moen 1976, 1978); black-tailed deer, *O. hemionus columbianus* (Wood *et al.* 1962), and reindeer, *Rangifer tarandus* (McEwan and Whitehead 1970, Leader-Williams and Ricketts 1982)) have been shown to be independent of resource availability. 'Seasonal' cycles in appetite, metabolic activity and body weight are still apparent even in captive animals kept at constant temperature and offered ad lib. food, but subject to the changing daylight regimes of the temperate year (e.g. McEwen *et al.* 1957, Kay 1979, Kay and Staines 1981, Suttie and Simpson 1985 – see also Putman 1988).

At least among the various ruminants studied therefore it would appear that the seasonal pattern of condition gain through late spring and summer, and condition loss through winter and early spring is a purely endogenous cycle entrained on daylength and associated with consistent changes in basal metabolic rate and appetite. Even when provided with unlimited food, captive animals still consume less over winter than is required to meet their physiological demands and in consequence suffer a depletion of body reserves.

Temperate feral horses show a similar cycle of condition to deer and other ruminants, and the question thus arises of whether the horse's cycle of body condition is also endogenously controlled or genuinely reflects a response to resource limitation. Available evidence is limited and contradictory. Pratt *et al.* (1986) showed that time spent feeding by New Forest ponies was less during winter than during spring and summer, and Kaseda (1983) also records that feral Japanese Misaki horses also spent less time feeding in winter, accompanied by a fall in metabolic rate. Ellis (1975) also showed that New Forest and Welsh Mountain fillies kept under normal photoperiodic conditions over winter and fed a low plane diet suffered a decrease in basal metabolic rate (BMR); however, this decrease in BMR appeared to be a direct response to undernourishment, since equivalent animals fed on a higher nutritional plane maintained a constant BMR and continued to gain body weight throughout the winter period. Further, other studies of free-ranging populations (Tyler 1972, Welsh 1975, Berger 1986) have shown that feeding rates and time spent feeding actually increase, or at least do not decrease, during winter, suggesting that any decrease recorded in other studies must be seen as an attempt to reduce energy expenditure at a time of low forage availability (when costs of extending foraging time are not balanced by equivalent energy returns from food gained), rather than the expression of some endogenous cycle in appetite and BMR.

In a detailed study of condition changes in different individual New Forest ponies over a 3-year period, Gill (1988; see also Burton 1992) showed that all experienced a similar cycle of condition change, with best condition occurring in late summer and autumn and poorest condition apparent in early spring, but that there were significant differences both in the absolute condition achieved by different ponies in late summer and the magnitude of condition loss through the winter (i.e. the difference between peak and poorest condition in any one individual) between individuals. Differences between individuals in peak condition reached and extent of condition loss over winter were explained in part by age and reproductive condition (lactating animals reached lower peak condition scores in autumn, and lost condition more rapidly through the subsequent winter; Gill 1988), as well as by parasitic infestation (intestinal nematode burden; Burton 1992). But a major part of the observed differences in both absolute condition and rate of change of condition could be attributed to differences between individuals in availability and use of resources. Thus significant differences in condition were apparent between populations of different areas of the Forest, characterized by different availability of shelter or preferred forages; differences were also apparent amongst individuals of a common home range, who used the available resources of forage or shelter differently.

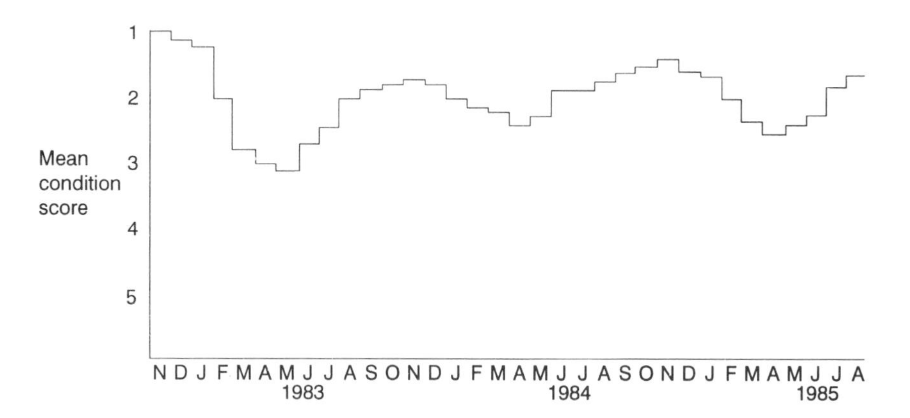

Figure 5.2 Seasonal cycles in body condition of New Forest ponies. The figure shows changes in the condition of mares and foals from November 1983 to August 1985, from data of Gill (1988). Condition is assessed by a visual index, which ranges between 1 (very good) and 5 (very poor), after Pollock (1980) and Gill (1988).

Animals that maintained consistently better condition during the year were found to make more use during both winter and spring of streamside lawns, wetland areas and deciduous woodland (especially that with much holly) than animals of generally poorer condition. In summer 'better' ponies were again found to use streamside lawns, deciduous woodlands and woodland glades more than animals of poorer overall condition, and ate greater amounts of both rushes and holly browse, while in autumn individuals found overall to maintain better condition were found to make significantly more use of habitat types offering both shelter and forage, such as deciduous woodlands and gorse (Gill 1988). Animals maintaining better condition were found to consume more rushes (*Juncus* spp.) and holly through winter, spring and summer, and more *Molinia caerulea* in spring. Ponies in relatively poorer condition were characteristically found to consume more *Agrostis curtisii* over winter and to spend more time out on open, exposed areas of dry heathland throughout the year. Similar correlations between individual patterns of resource use and body condition have since been demonstrated for free-ranging ponies on Exmoor and Dartmoor (Gill 1991) – with differences in condition associated with use of exactly the same forage types associated with better or worse condition within the New Forest.

Clearly, even if some physiological cycle of changing condition is endogenous in ponies as it is among ruminants, the **degree** of condition change varies and can be associated with differential patterns of resource use, suggesting that at least some part of the observed condition changes are due to deficiencies/limitation of resources. Manipulation of resource availability – and observation of observed changes in condition – provides a further test of the degree to which the cycle of condition is endogenous, or reflects genuine limitation of resources. In Gill's studies, better condition was associated with increased consumption of holly through spring and summer. In late winter 1988 (March) and again in 1989 (February) the availability of holly browse was deliberately increased by pollarding a number of mature trees in two areas of the Forest. Pollarding increases the amount of holly available for consumption by cutting down previously inaccessible high branches, which usually bear a higher proportion of the preferred smooth, 'blind' leaves without spines, leaving these cut branches accessible at ground level. The condition of animals observed feeding on the pollarded holly was compared against animals in the same areas not observed at the cut branches (Gill 1991); in both years the condition of animals attracted to the holly browse was significantly higher during March than that of animals not observed feeding on the pollard (although differences did not persist into April or May when condition of both groups of ponies was in any case increasing more generally).

Such evidence confirms some role of resource availability in effecting changes in condition. Gill also provides other anecdotal evidence that variation in condition reflects genuine changes in resource quality and quantity – even to the extent of suggesting that such changes may underlie the entire annual cycle observed and that there may be no endogenous component. A number of individual ponies running on the Forest are semi-tame and regularly hand fed by their owners during the winter. Records kept of the condition of these ponies showed that most lost little if any condition; young animals continued to grow and even lactating adults remained in fairly good condition. Unlike experimental ruminants offered unlimited food during the winter, these ponies took full advantage of supplementary foods offered and increased their intake accordingly – in the majority of cases at least to maintenance level (Gill 1988).

Such results are quite consistent with the very inconsistency of earlier studies (with some authors showing increases in time spent feeding and intake rates, others showing decreases in feeding rates over winter). If cycles in body condition are determined largely by simple abundance of resources over summer and autumn, and limited availability or quality of resources over winter, then animals in areas where sufficient resources *are* still available can compensate for lower availability in the environment or reduced nutrient quality by increasing foraging time; animals in areas where resources are simply insufficient to match requirement may opt instead for a strategy of energy conservation by reducing energy expenditure associated with active foraging in exposed areas.

5.6 FEEDING BEHAVIOUR OF CATTLE AND PONIES: DIFFERENT STRATEGIES OF EXPLOITATION

Clearly both cattle and ponies depastured on the Forest are preferential grazers – spending 50% or more of their time year round on the various improved grasslands of the Forest, and with diets of both species consisting largely of grasses (80–90% of intake in the summer for both species; still 50% or more through the winter).

But one of the most striking features of the data presented here is a marked difference in flexibility of patterns of habitat use and diet between the two species. Cattle show little flexibility of habitat use or diet and maintain essentially the same pattern of feeding behaviour throughout the year, relying on a very few forage species throughout: largely heather and grasses. By contrast, ponies utilize a far wider range of plant species at any one time and also show pronounced seasonal variation in resource use. Although they, too, concentrate on favoured grassland areas throughout the summer, and linger for some time after production is exhausted, they eventually drift towards exploitation of other forages during autumn and winter, when food becomes scarce. Increased intake of gorse and other browse materials compensates in some measure for the reduced availability of grass. It is difficult to account fully for this pronounced difference between the species, although Putman (1986b) suggests that partial explanation may be found in differences in the digestive physiology and social organization of the two species. Although both cattle and ponies are **preferential grazers**, the direct-throughput system of the monogastric horse does enable it, if necessary, to cope with alternative foodstuffs even if of low digestibility (Duncan 1992). By contrast, the whole design of the cattle rumeno-reticulum is geared for bulk fermentation of grasses (Hofmann 1982) and perhaps does not permit it to exploit efficiently a wider range of forage. Further, as noted earlier, the entire pattern of resource use of cattle is affected for those beasts that remain on the Forest during winter, by the provision of supplementary forage; it may well be that this provision of hay and straw in localized areas affects habitat use as well as the extent of natural foraging and contributes towards the lack of change in winter diet as observed for ponies.

But differences in foraging patterns may also be related to differing social structures. The basic social unit of ponies within the New Forest is the individual: a single mare, or mare plus current foal (Tyler 1972, Putman 1986b, Gill 1988). These individuals drift relatively independently around their home range, moving freely from one vegetation type to another. The cattle of the New Forest are, by contrast, strongly herding, and groups of between 30 and 40 individuals move as a unit around a defined range. The pattern of range use is almost stylized, with the herd following a fixed circuit each day, moving purposefully from one vegetation type to another, from night-time shelter in deciduous woods to daytime grazing areas. Further, social cohesion is such that they are unwilling to occupy vegetational communities whose extent is less than about 10 ha, since on smaller areas it may be impracticable to accommodate the entire herd. As a result, many of the vegetation types of the Forest which occur only in small patches are socially unavailable to them: the herd is restricted to woodlands, grasslands and heathlands.

Certainly, these differences in social structure, and the unwillingness of cattle to occupy communities on which the entire herd is not easily accommodated, leads to some spatial separation between the two species – and some difference in community use. Although during the summer, growing season, the pattern of habitat use for feeding and the actual diet are remarkably similar for the two herbivorous species, there is in practice some degree of segregation. Both concentrate their feeding on the various improved grasslands of the Forest, with peripheral use of wet and dry heathlands, acid grassland and bogs; in both species dietary composition at this time is over 80% grass. But although feeding use of grasslands by both species is extremely high at this time, they concentrate on different grassland types. Such separation is in fact apparent throughout the year and not just in this summer season of highest apparent overlap. Thus for example, cattle, preferring areas of greater than 10 ha in size, tend to concentrate most of their feeding activity throughout the year on reseeded grasslands (p. 76); other improved grasslands (areas of natural acid grassland, not reseeded but cleared of bracken and limed) are used during March, April and May – the peak growing season – while other smaller grassland types, such as streamside lawns or road verges, are rarely exploited. Ponies feed far more evenly over all grassland types, but none the less show a preferred use of the 'improved' grasslands (particularly from October to January/February), more consistent use of streamside lawns and relatively high use, compared to cattle, of roadside verges. Ponies also use two other grassland communities virtually never used by cattle – the truly natural acid grasslands of the Forest and the *Agrostis capillaris* swards of woodland clearings and glades.

Such observations indicate some measure of separation in feeding use of the different grassland communities by these two specialist herbivores, with some absolute differences (cattle do not use the woodland glades) and further differences in the importance of shared communities to the two species (cattle clearly exhibit very strong preference for reseeded areas, while ponies actively select limed grasslands, streamsides and road verges). In addition there is spatial segregation superimposed upon this: even for those communities used in common (improved grasslands and reseeded areas), cattle infrequently use patches smaller than 10 ha, which are thus occupied largely by ponies alone; on larger patches where both do occur, establishment of distinct grazing and latrine areas by the ponies, with cattle grazing restricted to latrines (p. 17) establishes effective separation on a local scale. But to what extent do these two herbivores compete? And what is the interaction between them and the various species of Forest deer?

6

The potential for competition

6.1 OVERLAPS IN RESOURCE USE

Even in describing the simple patterns of resource use of the different Forest herbivores, we have hinted at interaction. Thus both cattle and ponies are seen to be specialist grazers and, despite some separation between the two species in the classes of grassland exploited, there remains considerable overlap. Both spend 50% or more of their time throughout the year on the Forest grasslands (which in fact, at a total area of only some 1400 ha, represent a mere 7% of the total Open Forest area); both make up the bulk of their diet from the same few species of grasses. In the same way, Mann (1983) has suggested that at least part of the explanation for the striking difference in the diet of New Forest sika from that recorded for sika elsewhere in Britain, may well reflect lack of availability of grass forage within their range within the Forest because of prior grazing by the Forest ponies. Sharma (1994) notes that the diet of the Forest roe deer had changed significantly by the time of his studies in the late 1980s, from that reported by Jackson some 20 years earlier in the early 1970s (Jackson 1980); he suggests that while in part this may reflect changes in overall vegetational composition over the intervening time, due to growth and canopy closure within plantation conifer, reduced use of bramble, rose and field-layer forbs may well reflect a decreased availability due to prior exploitation by other herbivores.

From data presented in the previous chapters on patterns of habitat use and diet of the various species of large herbivores at free-range within the Forest, it is possible to calculate formal measures of the degree to which the different species overlap in their current use of resources. Within months, or groups of months selected to reflect biological seasons, overlap in the use of individual resources may be calculated using Pianka's (1973) index of overlap:

$$\propto_{j,k} = \frac{\Sigma p_{i,j} p_{i,k}}{[(\Sigma p_{i,j}^2)(\Sigma p_{i,k}^2)]^{1/2}}$$

(where $p_{i,j}$ and $p_{i,k}$ are the proportion made by the *i*th partition of a given resource dimension to total resource use by species j and k respectively). This index assumes values between 0 (total niche separation) and 1 (total overlap).

Putman (1986a,b) presented an analysis of the degree of niche overlap apparent in dietary composition between cattle, ponies, fallow, sika and roe deer, using data available at that time (cattle and ponies, Putman *et al.* 1987; fallow, Jackson 1974, 1977; sika, Mann 1983; roe, Jackson 1980). Overlap in observed patterns of habitat use was calculated amongst cattle, ponies, fallow and sika (data from Pratt *et al.* 1986, Jackson 1974, 1977, Parfitt unpublished data; sika, Mann 1983). Data were not available at that time on habitat use of roe, nor on habitat use or diet of red deer and neither species could be included fully in the analyses. Now, however, Payne (1987) has offered new data on habitat use and feeding behaviour of the Forest's red deer, and Sharma (1994) has presented a detailed analysis of habitat use by roe deer, as well as an updated overview of current diets of roe within the Forest. New data are also available on habitat use and diets of fallow deer (Thirgood 1990, 1995a, Putman *et al.* 1993) and ponies (Gill 1988).

Using these additional data we may attempt a review of that same 1986 analysis (Tables 6.1 and 6.2). The attentive reader will notice that all overlap values are changed, even those for which no new data have been provided here to revise or update resource use patterns presented (e.g. cattle with sika). For consistency with the seasonal definitions employed by Payne (1987), Sharma (1994) and Thirgood (1995a), all original figures have been recalculated for these analyses with a slightly different grouping of months than that employed by Putman (1986a,b). Thus winter is now, throughout, taken as December–February; spring, March–May; summer, June, July and August; and autumn, September–November inclusive.

Table 6.1 summarizes overlap indices calculated in relation to observed patterns of habitat use; Table 6.2 presents calculated overlap in relation to diet. Throughout the year, it is apparent from such analysis that there is relatively high overlap between cattle and ponies with respect to their use of habitat ($\alpha = 0.78$ overall), although it is at its highest in the spring and summer when both species are making extensive use of the favoured grasslands. Sika and fallow deer also show high overlap in habitat use throughout the year ($\alpha = 0.81$ for the year as a whole) and, particularly during autumn and winter, high overlap is also apparent between fallow and roe deer. However, it is in regard to diet that the

Table 6.1 Niche overlap between the various herbivores of the New Forest in relation to use of habitat. Overlap values (between 0, total separation, and 1 complete overlap) are calculated by the index of Pianka (1973), as

$$\text{Overlap } \propto_{j,k} = \frac{\Sigma\, p_{i,j}\, p_{i,k}}{[(\Sigma p_{i,j}^{2})(\Sigma p_{i,k}^{2})]^{1/2}}$$

		Cattle	*Ponies*	*Fallow*	*Sika*	*Roe*	*Red*
	Cattle	*					
	Ponies	0.44	*				
Winter	Fallow	0.43	0.46	*			
(December–February)	Sika	0.23	0.25	0.89	*		
	Roe	0.14	0.17	0.80	0.66	*	
	Red	0.08	0.22	0.12	0.02	0.26	*
	Cattle	*					
	Ponies	0.96	*				
Spring	Fallow	0.67	0.61	*			
(March–May)	Sika	0.19	0.14	0.79	*		
	Roe	0.10	0.07	0.61	0.50	*	
	Red	0.64	0.64	0.30	0.01	0.01	*
	Cattle	*					
	Ponies	0.72	*				
Summer	Fallow	0.41	0.52	*			
(June–August)	Sika	0.03	0.10	0.64	*		
	Roe	0.02	0.10	0.71	0.65	*	
	Red	0.42	0.50	0.21	0.05	0.08	*
	Cattle	*					
	Ponies	0.76	*				
Autumn	Fallow	0.31	0.33	*			
(September–November)	Sika	0.16	0.19	0.92	*		
	Roe	0.15	0.16	0.87	0.81	*	
	Red	0.20	0.34	0.22	0.23	0.21	*

Data used in these calculations are:
for cattle and ponies, data of Pratt *et al.* (1986), here Tables 5.1 and 5.2;
for fallow deer, data of Thirgood (1995a), here Table 3.6;
for sika deer, data of Mann (1983), Putman and Mann (1989a), here Table 4.1;
for roe deer, data of Sharma (1994), here Table 4.7;
for red deer, data of Payne (1987), here Table 4.4.

most significant overlaps in resource use between species are recorded (Table 6.2).

Dietary overlap between cattle and ponies is again high (0.9 or above in all seasons), despite the high intake of *Molinia* by ponies over the spring and summer (a forage not exploited at all by the Forest cattle) and despite their increased intake of browse materials such as holly and gorse during the winter months. The heavy reliance on grazing by both species still results in high overlap in this case – and is also responsible for significant overlap with the intermediate or bulk-feeding deer species: sika, red and fallow, although this, too, is at its most intense from March to July – over the main part of the growing season when food is relatively more abundant. There is throughout, far less overlap resolved

Table 6.2 Niche overlap between the various herbivores of the New Forest in food use. Overlap values are calculated by the index of Pianka (1973), as in Table 6.1 with 0 = no overlap, 1 = total overlap in resource use.

		Cattle	*Ponies*	*Fallow*	*Sika*	*Roe*	*Red*
	Cattle	*					
	Ponies	0.86	*				
Winter	Fallow	0.72	0.73	*			
(December–February)	Sika	0.75	0.67	0.89	*		
	Roe	0.17	0.16	0.66	0.49	*	
	Red	[0.29]	[0.16]	[0.65]	[0.73]	[0.47]	*
	Cattle	*					
	Ponies	0.95	*				
Spring	Fallow	0.98	0.97	*			
(March–May)	Sika	0.80	0.71	0.79	*		
	Roe	0.22	0.18	0.33	0.48	*	
	Red	[0.90]	[0.82]	[0.90]	[0.95]	[0.42]	*
	Cattle	*					
	Ponies	0.94	*				
Summer	Fallow	0.95	0.91	*			
June–August)	Sika	0.86	0.74	0.82	*		
	Roe	0.21	0.19	0.40	0.31	*	
	Red	[0.75]	[0.64]	[0.77]	[0.96]	[0.39]	*
	Cattle	*					
	Ponies	0.96	*				
Autumn	Fallow	0.79	0.80	*			
(September–November)	Sika	0.85	0.74	0.77	*		
	Roe	0.24	0.22	0.55	0.40	*	
	Red	[0.77]	[0.61]	[0.57]	[0.92]	[0.32]	*

Data used in these calculations are:
for cattle, data of Putman *et al.* (1987), here Table 5.4;
for ponies, data of Putman *et al.* (1987) and Gill (1988), here Tables 5.5 and 5.6;
for fallow deer, data of Jackson (1977), here Table 3.7;
for sika deer, data of Mann (1983), Putman and Mann 1989b, here Table 4.3;
for roe deer, data of Jackson (1980), here Table 4.5.

Data for red deer (Payne 1987) differ in form from those for the other species, being based not on ruminal or faecal analyses but on simple feeding observations.

between the diet of the Common stock and that of the concentrate-selecting roe.

Among the deer themselves, diets of fallow and sika deer show significant overlap throughout the year: both species are intermediate feeders by Hofmann's (1985) classification and clearly both select the same types of food. Diets of both species also show extensive overlap with that suggested for red deer – but we should treat these figures with some caution, in that data available for red are based on feeding observations only, while those for other species derive from direct faecal analyses. Overlaps in diet recorded with roe are also far higher for all three species (as mixed feeders) than those recorded between roe and the true grazing stock: cattle and ponies.

In calculation of overlap indices of Table 6.2, in respect of food use, the original data of Jackson (1977, 1980) have been used for fallow and roe, since sample sizes were larger than those on which later analyses of Putman *et al.* (1993), or Sharma (1994) are based – and data are in addition more contemporaneous with those used for the other species (cattle, ponies, sika). We might note, however, that equivalent indices can be calculated substituting these more recent data. Sharma (1994) notes that New Forest roe seem currently to be eating more coniferous browse as well as increasing their intake of heather; his analyses also suggest that the animals are able to rely far less extensively on bramble than at the time of Jackson's studies, and instead take significant quantities of ferns during late summer and autumn. If overlap values are recalculated using Sharma's results in place of those of Jackson (1980), overlap between roe and both sika and red deer is seen to be slightly higher than those presented in Table 6.2. [Overlaps with sika (original figures of Table 6.2 in brackets, for reference) are estimated at 0.62 (0.49), 0.50 (0.48), 0.52 (0.31) and 0.63 (0.40) in winter, spring, summer and autumn, respectively; overlap with red as 0.65 (0.47), 0.51 (0.42), 0.70 (0.39) and 0.58 (0.32).] Estimated overlap with fallow is not markedly altered by using data of Sharma rather than Jackson [0.57 (0.66), 0.25 (0.33), 0.55 (0.40), 0.50 (0.55)]. Jackson (1980) also noted that the diets of roe and fallow deer within the Forest showed greatest overlap in winter and early spring. Concentrating on this period of the year on the assumption that if there is any competition for food between the two species, it is likely to be at its most intense at this time, when food is shortest, he none the less concluded that widespread competition is unlikely to occur.

Such high overlaps in diet are striking – but should not be viewed in isolation. Apparent overlap on any single resource dimension may be far lower in practice when other dimensions of the niche are taken into account. The same foodstuff may (and in the Forest, commonly does) grow in a variety of different habitats: apparently high overlap in use of that resource may thus be resolved if the animals are in fact foraging for it in different habitats. True niche relationships are best examined by combining overlaps in individual resource dimensions into a multidimensional whole. Resources of habitat and food may be considered effectively independent; multidimensional overlap can thus be calculated in each case as the product of unidimensional values (May 1976, Pianka 1976, but see Putman (1994) for a critique). Table 6.3 shows combined niche overlap between the six ungulate species when both use of habitat and actual diet are considered simultaneously.

Table 6.3 Combined niche overlap between the various herbivores of the New Forest; figures derive as the product of separate overlap indices calculated in Tables 6.1 and 6.2 for habitat use and diet. 0 = No overlap, 1 = total overlap in resource use.

		Cattle	*Ponies*	*Fallow*	*Sika*	*Roe*	*Red*
	Cattle	*					
	Ponies	0.38	*				
Winter	Fallow	0.31	0.33	*			
(December–February)	Sika	0.17	0.16	0.80	*		
	Roe	0.02	0.03	0.53	0.32	*	
	Red	0.02	0.03	0.08	0.01	0.12	*
	Cattle	*					
	Ponies	0.91	*				
Spring	Fallow	0.65	0.59	*			
(March–May)	Sika	0.15	0.10	0.62	*		
	Roe	0.02	0.01	0.20	0.24	*	
	Red	0.58	0.53	0.27	0.01	0.01	*
	Cattle	*					
	Ponies	0.68	*				
Summer	Fallow	0.39	0.47	*			
(June–August)	Sika	0.02	0.07	0.53	*		
	Roe	0.00	0.02	0.28	0.20	*	
	Red	0.31	0.32	0.16	0.05	0.03	*
	Cattle	*					
	Ponies	0.73	*				
Autumn	Fallow	0.25	0.26	*			
(September–November)	Sika	0.13	0.14	0.71	*		
	Roe	0.04	0.04	0.48	0.32	*	
	Red	0.15	0.21	0.12	0.21	0.07	*

Combined overlaps can be seen to be significantly reduced, with differences in use of habitat compensating for the high dietary overlaps recorded between some species. Niche overlaps fall to levels generally below 0.5 (with 0.54 considered by some authors the critical limit to similarity for coexisting species; MacArthur and Levins 1967, Putman 1994). The highest remaining overlaps are seen between cattle and ponies in spring, summer and autumn (α = 0.91, 0.68, 0.73, respectively) and between fallow and sika deer over autumn and winter (α = 0.71, 0.80). Overlaps between the domestic stock and the deer are resolved to lower levels.

Even these figures may be misleading. While Tables 6.1–6.3 summarize the **potential** for interaction between the different herbivores of the Forest, based on overlap in observed patterns of resource use, it is clear that in some instances no such interaction is in fact occurring (yet!). Thus high overlap is shown in patterns of resource use, for example of cattle and sika deer; but at the time of recording, the sika of the New Forest were restricted to one small area in the south of the Forest, largely within a statutory Inclosure from which cattle, if not ponies, were excluded. Despite similarities in resource use, there was in practice no actual effect of that theoretical overlap, due to spatial separation between the species. The same argument belies the potential impact of observed overlap between sika and fallow or sika and red deer, for once again, patterns of resource use by sika were resolved in areas where neither red nor fallow deer occur(red).

This does not merely imply that the high apparent overlap is of no consequence because the animals are geographically separated; in a sense that is of small consequence: sika are now extending their range within the Forest and already may be considered potentially interactive with all the other Forest herbivores. More to the point it also may present a distorted picture of the extent of overlap that would persist where sika occur in genuine sympatry with these other species – for under such circumstances it is possible that they might at once respond with some shift in resource use, reducing overlap.

While data employed in estimation of the extent of overlap in resource use between the different Forest herbivores are for the most part derived from areas where the animals are interacting, data included for sika stand apart. Resource use patterns of cattle and ponies presented by Putman *et al.* (1987), Pratt *et al.* (1986), Gill (1988) and summarized here in the calculation of niche overlaps (Tables 6.1 and 6.2) are taken from areas where the animals genuinely co-occur; patterns of habitat use and diet thus reflect actual, expressed resource use in interaction. Data on habitat use and diets of fallow deer (Jackson 1974, 1977, Thirgood 1995a, Putman *et al.* 1993) are derived in the main from studies of animals in Forest Inclosures; however, the studies of Thirgood (1995a,b) and Putman *et al.*

(1993) did include animals whose ranges extend on to the Open Forest, and ponies at least were regularly encountered within the Inclosures – thus such resource use profiles may be taken as, effectively, realized niches in areas of sympatry at least with the Common stock, although not in areas frequented by red or sika. Data of Payne for red deer (Payne 1987) are taken from the Open Forest, where red deer co-occur with fallow, roe and Common stock; Sharma's data for roe are derived from areas where they also overlap with cattle, ponies and fallow deer: such data for red and roe may thus also be considered as expressions of realized niche in areas of genuine interaction with at least fallow deer and the domestic grazers. But data for sika as an influence are absent from all such analyses of other species, and the data presented by Mann (Mann 1983, Mann and Putman 1989a,b) for sika deer themselves, were derived in a localized area of the Forest where other deer species are absent. Although it seems probable that patterns of resource use by sika have been affected by pony grazing within their range, at least in terms of their relationship to those for the other deer species, Mann's data must be seen as allopatric representations of patterns of resource use by sika.

We can resolve this difference in analysis at least in part: we do have some more recent data on habitat relationships among the deer in true interaction (Boxall 1990). Although neither cattle nor ponies occur in Boxall's study area, all four species of deer (red, fallow, sika and roe), co-occur in true sympatry. Boxall's data therefore allow us to determine the degree of overlap/separation observed at least between the different species of deer in genuine interaction. Data are presented for patterns of overlap/separation in relation to use of habitat only; sadly, facilities did not permit analysis of diet of the different species.

Roydon Wood (already described on pp. 51–2), though formerly within the original perambulation of the Forest, now lies across the Forest fence, immediately abutting the boundary of the 'official' Forest to the south. In terms of habitat composition, Roydon Wood still resembles the wider Forest: much of its 380 ha is mixed broad-leaved woodland (predominantly oak, with birch and beech); there are two substantial blocks of Scots pine and several smaller areas of pine and larch. There are extensive areas of both dry, *Calluna*-dominated, heathland and wetter areas dominated by *Erica tetralix* and *Molinia caerulea*, patchworked through the site, and approximately 65 ha of the area is abandoned agricultural land or unimproved pasture. Being nowadays outside the statutory Forest boundary, cattle and ponies are excluded, but the site supports populations of fallow, roe, red and sika. Boxall (1990) has assessed patterns of habitat use of both sexes of all species from regular transect walks throughout the site, between 1985 and 1988. Detailed patterns of habitat use determined need not concern us here. Levels of overlap calculated in relation to habitat use are, however, universally high (Table 6.4), with overlaps in excess of 0.66 recorded between all pairs of species in all seasons.

Table 6.4 Habitat use overlap amongst the deer of Roydon Woods, calculated from data of Boxall (1990); overlap values are calculated by the index of Pianka (1973), as in Table 6.1. 0 = No overlap, 1 = total overlap in resource use.

		Fallow	*Sika*	*Roe*	*Red*
Winter	Fallow	*			
(December–February)	Sika	0.76	*		
	Roe	0.98	0.70	*	
	Red	0.99	0.73	0.99	*
Spring	Fallow	*			
(March–May)	Sika	0.66	*		
	Roe	0.82	0.74	*	
	Red	0.99	0.64	0.82	*
Summer	Fallow	*			
(June–August)	Sika	0.82	*		
	Roe	0.89	0.89	*	
	Red	0.96	0.72	0.73	*
Autumn	Fallow	*			
(September–November)	Sika	0.78	*		
	Roe	0.93	0.75	*	
	Red	0.86	0.57	0.66	*

6.2 RESOURCE LIMITATION

Even despite such refinements, the data presented above simply offer an analysis of the extent of overlap of resource use of a guild of interactive species. Although indicative of a potential for competition, overlap in observed patterns of resource use, whether high or low, does not necessarily imply anything about levels of competition in practice. As we have already noted more generally, competition through prior exploitation of resources is only of significance if resources may be shown to be limiting. Both Wiens (1989) and de Boer and Prins (1990) include demonstration of limitation of resources in their essential criteria for inferring competitive interaction in practice.

It seems unlikely that resources in the heavily grazed New Forest system are not limited at least during some periods of the year. The pronounced cycle of body condition in the Forest ponies reported by Pollock (1980), Gill (1988) and Burton (1992) may or may not have an endogenous component; the fact remains that the extent of condition loss over winter varies between individuals and is influenced, as well as by age and parasite burden (p. 84) but also by reproductive status and individual patterns of resource use (Gill 1988). Lactating mares (with higher energy requirements) appear unable to maintain condition as animals not bearing the added burden of lactation; ponies using certain forage-types more extensively than their neighbours (either because of overall availability or individual differences in selectivity) can also maintain better condition. Such differences in the seasonal cycle in condition, even if not the overall cycle itself, are certainly suggestive of some measure of resource limitation.

Further evidence is available from studies of actual above-ground productivity – and herbivore offtake – of a number of grazing associations within the Forest (Putman *et al.* 1981, Putman 1986b). The total above-ground production was measured in a number of different communities within the Forest between 1977 and 1979; Table 6.5 summarizes these figures and calculated offtake by large herbivores. Productivity and offtake of available plant material were estimated within grasslands, heather and bog communities using temporary exclosures to prevent grazing. In this method the amount of forage material present per unit area (standing crop) is assessed by cutting and weighing samples both when the exclosures are erected and at the end of the exclosure period.

Production may be then calculated as:

production = final standing crop (inside enclosure) – initial standing crop (outside enclosure)

and offtake as:

offtake = final standing crop (inside enclosure) – final standing crop (outside enclosure).

The length of time for which the exclosure is left is clearly critical. Vegetation inside the pen is of necessity protected from grazing; yet the vegetation it is supposed to represent normally grows under a continuous grazing pressure. With the relief from grazing afforded by the exclosure (necessary for us to take any measurements at all), we cannot be sure that the pattern of growth of the ungrazed vegetation within accurately reflects the growth pattern of the vegetation outside under its normal regime of heavy grazing. Relief from grazing must therefore be for as short a period as possible, yet cannot be too brief or the growth of vegetation within the pen is too small to measure. In practice pens on grasslands and bogs were sampled on a 2-monthly rotation; heathlands were clipped every 6 months (Putman *et al.* 1981).

Table 6.5 Estimated annual production and offtake in different forage communities of the Forest (from Putman 1986b); figures presented are as g/m^2.

Vegetation type	*Growing season production* 1977/8	1978/9	*Annual offtake* 1977/8	1978/9	*Offtake as % production* 1977/8	1978/9
Natural acid grassland	190	158	132	250	69.5	158.2
Improved natural grassland	321	329	313	371	97.5	112.8
Reseeded areas	349	226	221	296	63.3	131.0
Streamside lawns	473	492	432	523	91.3	106.3
Bogs	549	–	475	–	86.5	–

Production has been measured as above ground production over the growing season, while offtake is measured for the year as a whole.

1978/9 figures for offtake in relation to productivity frequently exceed 100%. This is not an artefact: offtake during this period of very heavy grazing did indeed exceed production and standing crop declined during the year. Note, however, that measurement of offtake as loss of standing crop does attribute all such losses to animals, and in practice loss of standing crop may also include a component from vegetation senescence.

Some care is needed in interpreting the results of such grazing exclosures – but that said, it is clear from Table 6.5 that levels of offtake in the Forest grasslands are extraordinarily high, amounting to nearly 100% of above-ground production for the year as a whole. Even during the summer growing season, with figures for production and offtake calculated month by month, 80–100% of the above-ground production is removed immediately by grazing animals.

Figures such as these suggest that at least in certain communities of the Forest – and certainly those exploited by the grazing species (cattle, ponies, fallow, red and(?) sika deer) – resources are likely to be limiting.

6.3 THE POTENTIAL FOR COMPETITION

Although it seems unlikely that resources in the New Forest are not limited in supply at least for part of the year, and thus that competition is avoided through simple superabundance of resources, inference of competition from recorded indices of resource use overlap is still far from straightforward. While high overlap in observed patterns of resource use may indeed be indicative of potential competition where resources can genuinely be shown to be limited, low observed overlap does not equally imply lack of competitive interaction – in that one response to competition might be a shift in the pattern of resource use of one or other species away from the zone of contest. Schoener (1983) has suggested that in large herbivores (and perhaps amongst mammals more generally) species are not readily shifted as a response to competitive pressure. But other authors have regularly reported such shifts, and thus low observed overlap in the field might also in some circumstances be considered indicative of competition (e.g. de Boer and Prins 1990), at least where the species concerned show potential for competition through extensive overlap in the pre-interactive, fundamental niche.

This distinction between overlap in the potential, **fundamental** niche and overlap in the expressed, or **realized** pattern of resource use is critical. Overlap in fundamental niche is indicative of a potential for competition where resources become limiting – unless coexistence is facilitated by some other factor such as predation, an additional competitor for one, but not both members of a competing pair, environmental perturbation at frequent intervals, etc. (p. 4). High observed overlap in realized niche may then reflect the expression of that potential competition in practice, if resources can, at the same time, be shown to be in limited supply. High overlap in fundamental niche, accompanied by **low** overlap in the expressed pattern of resource utilization, while clearly indicating a response to potential competition, still does not imply that competition is actually experienced at the present time; niche shift may have been sufficient to accommodate both species without further interaction. Finally, low overlap in realized niche alone, without prior knowledge of the degree of overlap in the fundamental, pre-interactive niche, tells us nothing: for low overlap in practice may reflect simple fundamental differences in ecology such that the species would never have a potential to interact in the first place.

Overlap values of Tables 6.1–6.3 are thus the more difficult to interpret because they are based on data collected during the course of several independent studies of each of the different species in turn – studied in varying degrees of sympatry/allopatry with others with whom we now wish to assess niche overlap. Overlap values presented are based on a curious admixture of initial resource use patterns, representing neither true fundamental nor post-interactive niches. All we may deduce is that overlap levels, certainly in terms of food use, are generally surprisingly high, and certainly represent a potential for competition – if resources become limiting.

Interference interactions

Analyses of niche overlap summarized above are dominated by considerations of overlap in resource use: and thus in essence focus primarily on the potential for interaction through prior exploitation and depletion of resources available to others. Yet competition between animal species may not only involve effects of one species on another through reduction in the available supply of resources. Competition may also involve an element of interference: physical interference in freedom of access to those resources. Individuals of sympatric species may coincide in time and space in their attempts to exploit the same patch of resources, promoting some direct, behavioural interaction. In effect we may recognize within the context of competition two distinct types of interaction. Competition may be direct, through direct interactions of this sort between individuals seeking to use the same resource and thus interfering with each other's access to that resource, or it may be indirect, with one individual affecting the other's free use of the resources through prior consumption. Often a competitive interaction will contain elements of both interference and exploitation, but it is important to distinguish between the two different types of interaction for they may have somewhat different effects (Putman 1994, pp. 77–8). Thus separation in time or space reduces primarily the potential for interference interactions; separation or overlap in diet reflect more a potential for exploitation competition.

In this case, even extensive overlap in patterns of habitat use *per se* (Tables 6.1, 6.4) is immaterial in that the **habitat** is not 'consumed'; overlap becomes significant only where patterns of overlap in habitat combine with patterns of overlap in use of a genuine depletable resource, such as food, such that they exacerbate, or at least do not mitigate, overlap expressed in that consumptive interaction (Table 6.3) – or if the animals try to use the same patch of habitat **at the same time**, and thus affect each other's free use of that space by interference. Analyses of resource overlap presented so far focus implicitly on the potential for exploitation competition. Overlap in patterns of habitat use is considered only in terms of its effect in reducing/increasing the degree of overlap apparent in patterns of forage use. No analysis has been presented of the frequency of actual coincident use of habitat patches.

In Boxall's 4-year study of fallow, roe, sika and red deer in Roydon Wood, individuals of two species or more were recorded in the same sampling unit (1 ha grid square) only on 30 occasions in a total of 2580 observations (Boxall 1990); however, animals of different species were commonly recorded within different patches of the same habitat, and there was no evidence of interference to exclusion. Similar conclusions are reported by Sharma (1994) who examined the potential for direct interaction between fallow and roe deer where they occur in sympatry, assessing from transect data the expected number of transect walks on which no deer were seen of either species, the number on which only roe, or only fallow were seen, and the number in which both roe and fallow might be encountered within the same site. Observed frequencies were compared against these expected values (using Chi-squared test for goodness of fit); no tests showed a significant difference between observed and expected values, offering no evidence that either species makes any attempt to avoid a site because of the presence of the other (Sharma 1994).

Perhaps the most obvious example we might cite of behavioural interaction is local separation that may be observed between cattle and ponies grazing common pasturage. As noted earlier (p. 17), free-ranging ponies of the New Forest, like their more domesticated counterparts in fields, establish within their grazing grounds distinct and traditional sites for grazing and for elimination (Edwards and Hollis 1982). The animals crop close to the ground in areas selected for grazing, but do not themselves graze within the areas established as latrines, except in occasional periods when other forage is extremely scarce. Although cattle (and some deer) also use the same areas – either simultaneously with the ponies or sequentially – they are unable to feed effectively in areas grazed by ponies because the ponies can crop the sward so close; feeding cattle or deer are forced to concentrate their foraging within pony latrine areas where the sward is longer (p. 88). In effect, the grazing style of the ponies excludes the cattle from certain patches even within a single feeding ground – though whether this should be interpreted strictly as an interference interaction, or rather an extreme expression of prior exploitation, is perhaps open to conjecture!

6.4 ECOLOGICAL INTERACTION AND POPULATION TREND

Overlap in resource use, interference, resource use by one species reduces its availability to another species . . . one of the most crucial of Wiens' (1989) criteria for detection of competition in the field is that one or more species can be shown to be negatively affected by the interaction.

Changes in numbers of cattle and ponies on the Forest are determined primarily by purely economic considerations. However, as noted in Chapter 2, in order to monitor changing size of the deer populations within the Forest, and to plan any necessary cull to control numbers or limit timber damage, the Forestry Commission has, since 1960, undertaken a regular annual census of the number of deer present of each species within the Forest's enclosed woodlands. Fluctuations in number of each species and long-term change may thus be examined in relation to the changing abundance of other large herbivores within the Forest over the same period (Putman and Sharma 1987).

The annual census undertaken by the Forestry Commission is designed to monitor trends in population rather than assess actual number. The keeper responsible for each of the 12 Forest 'beats' makes a return to the central office each year of the number of deer of each species held within his beat on a fixed date in April. In practice (p. 27) this figure is reached as a compromise between the keeper's accumulated knowledge of the number of deer in the area over the previous few weeks and an actual 'single-day' census on the appointed date. Counts relate to Forestry Inclosures only, but since most of the deer are concentrated in such areas, they do approximate to a census for the whole effective Forest area. More to the point, any such visual count consistently underestimates population size in concealing habitats, and we have already noted that population sizes for fallow, sika and even red are probably substantially higher than those recorded in the census. The population estimates returned are thus not (nor are they assumed to be) absolute values. However, as a relative index of population change from year to year (p. 66), the census figures may offer a reliable measure of overall trend.

Putman and Sharma (1987) have considered changes in the abundance of each of the herbivore species over a 24-year period in relation to the changing abundance of the various other ungulates, and changes in the Forest's vegetational cover. Treating each of the individual beats as replicates, correlation analyses considered for each species the censused population in any year, in relation to numbers in each beat of roe, fallow, sika and red deer in both the same, and the previous year; numbers of cattle, ponies and pigs depastured (or at least licensed) on the Forest in the same year, and a rolling 3-year mean of the number of cattle and ponies pastured on the Forest in the previous 3-year period; finally, censused numbers in each deer species were also related to a variety of vegetational descriptors (Putman and Sharma 1987, and above, p. 66), but these need not immediately concern us here.

No consistent trends in population number of the different deer species of the Forest are immediately apparent from the census figures (Table 2.1, repeated here as Table 6.6), except in the case of sika deer, for which the census estimates – and thus perhaps actual population numbers – have steadily increased since 1962. Red deer populations were clearly extremely sparse in the early 1960s, if not actually extinct; reintroductions were made in 1962 and populations appear now to have stabilized at a census estimate of between 60 and 70 individuals. If we believe the census as a consistent index, roe and fallow numbers in 1985 did not differ greatly from those recorded in 1962, although both populations show evidence of fluctuation and, in the case of the roe deer, these fluctuations are fairly wide – with a peak in 1969–70 that was more than double levels of 1962 or 1983–85.

However, changes in the numbers of each of the different deer populations appeared from correlation analysis to be, for the most part, independent of one another, suggesting little interaction between the species at all. Putman and Sharma (1987) noted, however, that numbers of roe deer over that 24-year period showed significant negative correlation with grazing pressure imposed upon the Forest by domestic stock. Roe deer numbers in any year were found to correlate with the mean

Table 6.6 Changes in censused numbers of New Forest deer populations 1960–92.

	Red	*Sika*	*Fallow*	*Roe*
1960	–	35	788	211
1961	–	44	857	302
1962	7	34	931	328
1963	23	42	933	383
1964	20	64	994	434
1965	23	84	1147	465
1966	24	97	1036	490
1967	29	67	968	527
1968	33	38	855	539
1969	18	38	893	572
1970	23	37	1022	620
1971	26	71	866	444
1972	28	73	1017	428
1973	29	75	904	435
1974	37	78	909	356
1975	49	88	834	375
1976	43	94	973	344
1977	49	80	908	363
1978	79	79	927	332
1979	34	80	950	354
1980	58	74	1016	334
1981	73	95	996	309
1982	64	76	1049	336
1983	72	97	1020	281
1984	57	115	1033	264
1985	66	99	1006	260
1986	70	103	1086	265
1987	77	103	1227	249
1988	77	79	1152	265
1989	94	90	1280	255
1990	145	78	1180	295
1991	97	101	1204	351
1992	93	96	1290	369

This table duplicates Table 2.1 in presenting censused numbers of the different species of deer in the New Forest based on the annual census carried out by the Forestry Commission, offering an index of population trends from 1960 to 1992.

numbers of cattle and ponies pastured on the Forest in the preceding 3 years ($r = -0.47$; $P = 0.02$; Putman and Sharma 1987).

Sharma (1994) has more recently updated this analysis of the effects on ungulate population numbers of the abundance of other herbivores within the Forest, and vegetational change – and has extended the analysis to explore in addition any possible effects of climatic factors on observed population change. Sharma's re-analysis is based only on data from (and including) 1972, although the data now extend until 1988 and thus still span 17 years. On close examination of the data from the early years of the census, Sharma expresses reservations about its reliability and consistency. As he notes, the idea of a regular annual census was only initiated in 1960 and, as a new and developing technique, it may not have been applied consistently from one year to the next (Sharma 1994).

At the end of the 1960s, a number of changes were made in various aspects of the Commission's range administration. New head keepers were appointed and the census was applied consistently to all areas of the Forest using the same or similar methods; in addition the head keepers attempted to apply some form of 'quality control' check of figures returned. Further, over the entire period from 1972 to 1986 (or with only one keeper change, to 1988) the same individual keepers remained within the same management 'beats'; the census data from 1972 to 1988 at least may thus be taken as a fairly reliable index of population change from year to year, since they have over that period relied on the same method and the same observers within each beat.

Fallow deer numbers over the whole Forest from 1972 to 1988 correlated significantly negatively with numbers of cattle and ponies pastured on the Forest in the previous year (cattle individually: $r = -0.53$; ponies individually: $r = -0.49$; or both combined: $r = -0.62$; $P < 0.05$ in all cases); fallow numbers were also significantly negatively correlated with the accumulated grazing pressure from domestic stock over the previous 3-year period ($r = -0.59$; $P < 0.05$). Censused numbers of fallow also showed a significant negative correlation with numbers of roe deer; indeed, estimated numbers of roe were found to show a significant negative correlation with numbers of all other deer species in the same year: with fallow ($r = -0.68$), red ($r = -0.82$) and sika ($r = -0.72$; all $P < 0.01$). But in Sharma's analysis of the period 1972–88, no significant correlations emerged between recorded numbers of roe deer and numbers of cattle or ponies, individually or in combination.

Correlations are that and nothing more – they cannot impute causality. We know that the number of red and sika have been increasing steadily in the Forest since the early 1960s; we know that roe have been declining at least since 1972. But both red deer and sika deer have, until recently, been extremely restricted in their distribution within the Forest (in each case, effectively restricted to occurrence within only a single beat) and it seems rather unlikely that such localized populations can have affected the total population of roe over the Forest as a whole. However, fallow are more widespread; their numbers, too, have been increasing as populations of roe have declined. However, the negative correlation established over the Forest as a whole is not maintained when examined at the level of individual beats; at this level, no consistent significant correlations between roe and fallow numbers were determined – suggesting that at the local level (where interaction might be expected) fallow numbers had no significant impact on roe deer populations, and thus the correlation observed over the Forest as a whole is due merely to coincident but independent increase in fallow numbers and decline in roe.

Indeed, roe population size was found in Sharma's analyses (Chapter 5) to show far stronger association with changes in vegetational composition and structure within the Forest, than with population size of any of the other ungulates. In multivariate analyses, undertaken to explore the relative importance of all the various environmental factors found to account for variation in censused roe populations, habitat variables alone (and notably the available area of prethicket conifer, before canopy closure) accounted for 93% of the recorded variation in estimated roe populations, with climatic factors contributing only a further 4% to explained variation (Sharma 1994, Putman *et al.* 1996).

While there may be limited evidence that grazing pressure from cattle and ponies deleteriously affects populations of fallow (Sharma 1994), and perhaps roe (at least suggested by the analyses of Putman and Sharma over the longer period of 1962–85), there is no consistent evidence for interaction among the populations of the different deer species themselves. But one additional problem as yet unaddressed is the extent to which population changes are determined more by human intervention than by any environmental characteristic. None of the populations studied is in practice unrestricted; all are subjected to a regular cull. It is possible that the dynamics are more determined by the level of the annual cull, or that numbers of all species are maintained through this imposed mortality at levels below those at which potential competitive interaction would be expressed (p. 4).

For the past 30 years or so, the Forestry Commission have established cull levels for each of the deer species within the Forest in a deliberate attempt to maintain stable populations within levels compatible with commercial forestry. On the assumption that two-thirds of the censused population of adult females (1 year and over) will produce offspring in any year, in a 1:1 sex ratio, an estimate is made of the number of individuals to be recruited to the population in that year. Estimated losses due to poaching and road traffic accidents are then deducted and additional cull required to maintain populations at constant level is calculated. Total imposed mortality over the period from 1972 to 1988 (from accidental death, poaching and culling operations) has been of the order of 40–50% of the censused population for fallow and sika deer, which might be assumed to have a major impact on population levels; culls of red deer have also been increased of late.

However, such a calculation assumes that the censused population in April approximates to the true population total, and that the mortalities to be expected from poaching and road traffic accidents can be assessed accurately. In his studies of population change of New Forest fallow deer, Strange (1976) considered that estimates for road traffic and poaching losses may well be reasonable. But we have strenuously avoided here any suggestion that the census figure on which the cull is first calculated may be an accurate measure of true population size. Visual censuses in wooded environments are renowned for underestimating actual population size (e.g. Andersen 1953, Staines and Ratcliffe 1987, Ratcliffe 1989), and we have already suggested in Chapter 2 that we consider the census figures a considerable underestimation of true population size: certainly for fallow and sika, probably also for red (see p. 27).

Strange's figures for 1970–75 suggest that the true population of fallow deer over that period was around 1800: almost exactly double estimates returned by the Commission census (a mean over that same 6-year period of 914 fallow deer); Mann's (1983) figures for sika deer populations between 1979 and 1982 suggest that populations must have numbered between 175 and 200: again double the census figure (74–95) for the same period. Population size for red deer is probably similarly underestimated (Payne 1987), although we have argued that it is probable that the census figure for roe more closely approximates to the true total (p. 27). On this basis, the true culling levels for the Forest deer populations over the period from 1972 to 1988 are more likely to have been in the region of 20–25% for fallow and sika deer, 12–15% for roe and red.

An imposed mortality of even 20–25% is certainly sufficient to have a disruptive effect on fallow and sika deer populations. Clutton-Brock *et al.* (1982) found that numbers of red deer on Rum increased substantially following the relaxation of a 17% cull in 1973; Ratcliffe (1987b, 1989) and Ratcliffe and Mayle (1992) have suggested that culls of between 18 and 22% are sufficient to control populations of red deer or roe deer in commercial forests, and Houston (1982) considered that human predation on wapiti populations in the Yellowstone National Park, at an average level of 21%, was the major factor responsible for changes in population number. It may thus be possible that culling in the New Forest is a major influence on population levels of all deer species, and may be one reason why Putman and Sharma (1987) and Sharma (1994) failed to find any consistent correlations between numbers of the different species suggestive of direct interaction.

7

Factors structuring resource relationships in ungulate assemblies

7.1 INTERACTIONS AMONGST NEW FOREST HERBIVORES

Grazing pressure in the New Forest is high and there is certainly no evidence of facilitation amongst the herbivore assembly. Indeed, pressure of grazing on the limited area of natural and improved grasslands is sufficient actually to cause a reduction in productivity. While low levels of offtake do appear in certain circumstances to result in a compensatory increase in growth rates, such that light grazing may appear actually to increase productivity (e.g. McNaughton 1979), heavy grazing pressure reduces the effective leaf area available for photosynthesis and depresses productivity. Grazing by cattle and ponies in particular removes from New Forest grasslands nearly all above-ground production as soon as it is produced, and Ekins (1989) has calculated that the pattern of grazing reduces production within the grassland to 0.45 of the maximum potential yield. There is no evidence from these Forest grasslands that any level of grazing stimulates productivity and thus facilitates grazing by other species; all levels of grazing **reduce** yield for others.

At the same time, there is evidence of clear and substantial overlap in resource use between many of the Forest's herbivores. Cattle and ponies clearly show very considerable overlap in diet through much of the year, and although there is some degree of segregation between the species in terms of the types of grassland favoured, there remains substantial overlap in habitat use. Formal analyses of niche overlap also suggest significant overlap in diet between the three mixed feeders among the deer (red, sika and fallow) and between these species and the domestic stock. De Boer and Prins (1990) argue that interspecific competition between any two species is only possible where:

1. there is clear evidence of habitat overlap;
2. there is overlap in forage consumed by the two species within those shared habitats; and
3. the shared dietary resources are limiting.

For many of the species of large herbivores of the New Forest, these conditions are met. There is resource overlap in both use of habitat and in preferred diet consistent with a potential for competition. Data presented in Table 6.5 suggest that at least for many of the grassland communities, grazing herbivores remove the entire above-ground production of digestible forage – and certainly the seasonal cycle of condition change amongst the Forest ponies (Gill 1988) is also consistent with clear resource limitation.

But de Boer and Prins present minimum conditions to offer even a potential for competition. Wiens' (1989) criteria for establishing the actuality of competitive interaction in practice further require that 'one or more species is negatively affected' and that 'observed patterns (of population trend or shifts in resource use between sympatry and allopatry) are consistent with predictions from competition'.

Mann (1983) has suggested that differences in the dietary composition of sika deer in the New Forest from those elsewhere within the UK are consistent with a reduction in the availability of suitable grasses within their range due to close-grazing of rides and glades by Forest ponies. Sharma (1994) also suggests that differences in the diet of New Forest roe recorded by himself in the late 1980s, by comparison to those reported by Jackson from studies between 1972 and 1974, are due to changes in the Forest vegetation, at least partly in response to continued grazing by other species as well as increasing maturity of forest plantations. And analyses by Putman and Sharma (1987) and Sharma (1994) of changes in population size of all deer species over a period of 25–30 years offer some evidence, perhaps, of negative effects of grazing by cattle and ponies on population densities of roe deer (Putman and Sharma 1987) and fallow (Sharma 1994).

All evidence presented so far would also suggest that there is probably a degree of actual competition between the cattle and ponies themselves, although it is hard to demonstrate this incontrovertibly. Populations on the Forest are not self-regulating, and changes in numbers recorded over the years probably reflect economic response to market demands to a greater extent than they show a response to changing resource availability; further, populations are managed so closely (with the provision where necessary of supplementary fodder, and removal from the Forest of any animals seen to be in poor condition) that any long-term effects of competition are masked by effects of husbandry.

Among the deer, observed overlap in patterns of resource use clearly reflect some **potential** for competition, but there is no clear evidence of interaction. Although it seems possible that there is some impact on fallow and roe populations from grazing of domestic stock, no evidence was found for any species of competition experienced from other species of deer. Both Putman and Sharma (1987) and Sharma (1994) show a significant negative correlation between numbers of fallow deer and roe deer within the Forest as a whole between 1960 (1972) and 1988, but such correlations were not maintained with analyses repeated at the level of individual Forest beats. Sharma's direct observation showed no significant positive or negative association between the two species and the correlation reported is more consistent with independent responses of both species to habitat changes over the period, favouring fallow but unfavourable to roe (Sharma 1994).

For none of the other deer species were trends in population size between 1960 and 1988 considered indicative of any effects of competition. However, as we have already suggested, such a result may be due to the fact that populations of deer in the New Forest, like those of the cattle and ponies, are managed to a degree and cannot be considered self-regulating. As noted above, the annual cull of all species taken by the Forestry Commission is sufficient to maintain populations at levels below those at which they might naturally equilibrate.

It seems entirely possible that lack of apparent competition amongst the 'wild' herbivores of the Forest is due to the fact that populations are maintained by artificial 'predation' to levels below those at which the particular resources for which they might compete amongst themselves become limiting. By contrast, populations of domestic stock are kept high by economic demand and artificial subsidy. In consequence, resources used extensively by cattle and ponies (grasslands and grasses) do become limiting; where overlap in resource use between the Forest's ungulates is in relation to these particular commodities (as between cattle and ponies, or cattle and ponies with fallow, red or sika deer) explicit competition may become apparent.

7.2 DOES COMPETITION OR PREDATION STRUCTURE UNGULATE ASSEMBLIES?

The question with which we ended the last section: the extent to which competitive relationships among New Forest ungulates may be masked or accommodated by independent control of population levels by predation (albeit in this case predation by human agencies) sounds many echoes. There has been a continuing debate amongst community ecologists over the years as to whether competition or predation is the more influential of biotic factors controlling structure (species number, population sizes, species interactions) and resource relationships in ecological communities in general (e.g. Salt 1984, Strong *et al.* 1984). That same question has been tackled explicitly amongst ungulate assemblies by Sinclair (1979, 1985).

While for the most part in this book we have focused primarily on temperate systems, any analysis of resource relationships among ungulates which passed by without reference to the classic series of studies from the Serengeti National Park in Tanzania would not only be guilty of gross discourtesy, but would also risk open ridicule, for the pioneering analyses of Sinclair and co-workers have in themselves largely shaped our current thinking. Although it suffers the dominating additional influence of seasonal migration, with some species of the community effectively major players only on a part-time basis, the rich ungulate assemblage of the Serengeti ecosystem offers in other ways striking parallels to our temperate assemblies – with one essential difference: the ten major species of ungulates within the Serengeti (wildebeest, zebra, buffalo, Thompson's and Grant's gazelles, topi, kongoni, impala, waterbuck and warthog) coexist in a system where natural predators still remain. Predation is without doubt a very significant pressure on ungulate populations: some 30% of all wildebeest mortality, for example, is due to predation, and predators account for between 60 and 74% of the annual mortality of adult zebra (Sinclair and Norton-Griffiths 1982).

Early descriptions of resource relationships amongst the ungulates of Africa concentrated for the most part in describing separation, by habitat (Lamprey 1963, Bell 1970, Jarman 1972, Sinclair 1977), by food species (Field 1968, 1972, Jarman 1971) or plant part (Gwynne and Bell 1968, Bell 1970, Sinclair 1977). The underlying presumption of such work was that niche separation was to be expected as the result of interspecific competition and as an adaptation to accommodate coexistence. In a more recent analysis of resource relationships amongst the ungulates of the Serengeti, however, Sinclair (1985) notes that although each species of ungulate had particular habitat preferences consistent with those reported by earlier authors, the more striking conclusion is the degree of **similarity** between such preferences in both wet and dry seasons, resulting in practice in extensive overlap in habitat use. Furthermore, there is also extensive overlap in forages selected: wildebeest, zebra, topi, buffalo and Thompson's gazelle eat the same plant components within the vegetational communities occupied, with resulting mean dietary overlap of 82.5% (Gwynne and Bell 1968, Sinclair 1977). Similar results are reported by Leuthold (1978) for browsing ungulates in Tsavo National Park in Kenya, with high overlap observed in both wet and dry seasons.

Since predation imposes such major mortality on populations of all species within the grazing assemblage, it is perhaps arguable that the herbivores may coexist despite extensive niche overlap, because numbers are maintained at levels below which resources become limiting. In this particular case, however, such an argument seems improbable, since the grazing herds indubitably remove most available vegetation; further, despite the heavy toll taken by predators, there is no good evidence to suggest that predation actually regulates numbers of the various ungulate species within the grazing assemblage (Schaller 1972, Sinclair 1977). Despite this, Sinclair argues that predation may still play a major role in structuring resource relationships. If risk of predation is high, there may be selective advantages in associating with other individuals, to the extent of forming mixed herds in open areas with other vulnerable species, even though such a grouping may be at the expense of increased competition for food (Sinclair 1985, de Boer and Prins 1990).

From his detailed analyses, Sinclair (1985) concludes that patterns of resource partitioning amongst the Serengeti's herbivores are consistent for the most part with avoidance of interspecific competition, but that in some cases, positive association is observed despite the increased potential for competition, in response to the protection afforded from predation risk. Sinclair notes that patterns of resource use suggest that Thompson's and Grant's gazelles are strongly influenced by predation risk and, despite the possible increase in interspecific competition that may result, positively seek association with the more vulnerable herds of wildebeest for the protection from predators that such association offers. Patterns of resource use by zebra, topi, impala, waterbuck and warthog are consistent with response both to predation risk and competition. Data for kongoni are equivocal: although they clearly do display some influence from predation risk, results suggest they may be more strongly influenced by actual feeding facilitation from wildebeest and competition from other species.

Positive facilitation is hard to prove (de Boer and Prins 1990, and Chapter 1 here) because of the difficulties of interpreting correlative data. Equally, the implication of interspecific competition as a structuring force solely by inference from recorded patterns of overlap or separation in resource use is likewise fraught with problems. As noted on p. 8, such circumstantial evidence only gathers conviction if actual changes in the expressed pattern of resource use are apparent between sympatry and allopatry, and/or if high observed overlap is associated with long-term changes in population dynamics of the species concerned.

In the New Forest assembly, no consistent changes in population number of any species of deer could be related to changing abundance of any of the other wild species, although, there is perhaps some evidence to suggest that changes in the numbers of domestic grazers (cattle and ponies) may have negative effects on population densities of roe deer and fallow (Putman and Sharma 1987, Sharma 1994, and see below). However, Chapman *et al.*(1985) recorded high overlap in diet and winter habitat use of roe deer and Chinese muntjac in the King's Forest in Suffolk, UK (Chapman *et al.* 1985, Forde 1989) and suggested that there is increasing evidence of a decline in numbers of roe in areas of high muntjac density (Wray 1994).

In this mixed woodland area patterns of habitat use by muntjac and roe deer showed significant overlap. The King's Forest is a mainly coniferous forestry plantation, largely planted with Corsican pine (*Pinus nigra*) with Scots pine, European larch and Douglas fir. However, many of the older broad-leaved coverts and shelter-belts remain within the mosaic, and amenity fringes of broad-leaves have also been added more recently. Muntjac in the King's Forest preferentially selected areas of broad-leaved woodland with a well-developed understorey of bramble (*Rubus* agg.); muntjac distribution was strongly correlated with both abundance and distribution of bramble (Wray 1994). Roe were also found preferentially in areas with a high abundance of bramble, and patterns of habitat use showed high overlap at all times of year (October, 0.90; April, 0.84; in the same periods calculated overlap in habitat use with sympatric fallow deer was 0.50 for muntjac (October and April), and for roe 0.56 and 0.67 in October and April, respectively; Wray 1992, 1994). The clear preference by both species for areas with good growth of bramble was also reflected in high overlap in diet: calculated dietary overlap was 0.89 over the year as a whole and in excess of 0.74 even in the season of lowest similarity (Wray 1992). No direct evidence for interaction has been observed in field studies, but in areas of high muntjac density there appears to be geographical displacement of roe (Chapman *et al.* 1993, Wray 1994).

Perhaps the most extensively documented temperate forest system is that of the Bialowieza primeval forest straddling the border between Poland and Belarus. Population estimates of European bison (*Bison bonasus*), moose (*Alces alces*) wild boar (*Sus scrofa*) red deer, roe deer and fallow deer have been attempted each year since 1890; these statistics are summarized by Jedrzejewska *et al.* (1996) in a detailed analysis of population trends. Whether conducted over the entire period for which data are available (1890–1993), or on the more robust data set for years since 1946, highly significant positive correlations are apparent between censused numbers of red deer, roe deer, boar and bison. Fallow deer (introduced in any case as an exotic in 1890) were extinct by 1920, but over that 30-year period, population numbers also showed significant positive correlation with numbers of red deer, roe deer, boar and bison. Moose numbers showed weakest correlation overall with those of other species (although, once more, significantly positively correlated with all species where analyses are restricted to 1946–1993). No significant negative correlations emerged between estimated populations of any species – and certainly no evidence for competitive interaction between them despite high potential overlap in diet and habitat use.

Jedrzejewska *et al.* (1996) conclude that the most important factors influencing absolute and relative population sizes of the different ungulate species of the Bialowieza community were long-term climatic trends causing shifts in forest structure towards an increasing proportional representation of broad-leaved trees (Dabrowski 1959, Jedrzejewska *et al.* 1994), resulting in a corresponding increase in available food supply; the effects of natural predators (Okarma *et al.* 1995, Jedrzejewski *et al.* 1996) and exploitation (and promotion) by man (Jedrzejewska *et al.* 1994); and the short-term effects of severe winter weather (e.g. Jedrzejewski *et al.* 1992).

We might note that forest structure, winter weather conditions and imposed cull mortality were also the factors recorded by Sharma (1994) as being the major factors responsible for recorded variation in censused numbers of individual species (fallow and roe) within the New Forest, and the Bialowieza data offer an elegant corroboration of our earlier suggestion that the population dynamics of at least the wild species of ungulates (red deer, sika, fallow and roe) were relatively independent of one another and could better be explained in terms of factors affecting each in isolation than by any interaction between them.

7.3 THE EVIDENCE FOR COMPETITIVE INTERACTION IN NATURAL UNGULATE ASSEMBLIES

In fact, in general, evidence for actual competition between members of an established guild of large herbivores is hard to find. From all the examples we have cited here and in our introduction (pp. 7–8) we find little incontrovertible evidence for competition as a dominating force influencing population dynamics or resource relationships within ungulate assemblies. This absence of reported competition may in part reflect lack of appropriate investigation, but is perhaps, in any case, not unexpected. Over an evolutionary time scale, natural selection would be expected to promote clear separation in resource use between regularly interacting sets of species specifically to minimize the loss of fitness incurred through competition. In effect we might expect that competitive interactions would become apparent only when an established system is perturbed from equilibrium in some way – perhaps challenged by a recent invader. In such context it is noteworthy that Chinese muntjac are an introduced species in the UK and were first recorded in the King's Forest in 1963; the potential for competition described by Chapman *et al.* (1993) and Wray (1992, 1994) may thus reflect the fact that muntjac and roe deer have not shared a long common evolutionary history. Likewise, sika deer, considered possible competitors for white-tailed deer on Assateague Island in Maryland, US, were introduced to the island in 1925 (Keiper 1985); and the most widely quoted examples of competition within an ungulate community report on interactions in New Zealand between red deer and sika (McKelvey 1959, Kiddie 1962), and between red deer, fallow and white-tailed deer (Kean 1959) – where the entire ungulate assemblage is introduced (Challies 1985).

Indeed, those few studies that have demonstrated clear population interactions, or have demonstrated niche shift, almost all seem to derive from situations where the system is perturbed by introduction of a new 'exotic' or where additional grazing pressure from domestic livestock is imposed upon a natural ecosystem. There are instances here, too, of potential facilitation: for example, summer grazing by cattle was shown to increase the quality of grass swards available to feeding red deer hinds on the Isle of Rum (Gordon 1988) and Rhodes and Sharrow (1990) report that sheep grazing appeared to improve forage quality in autumn and spring for white-tailed deer and wapiti in Oregon, and increased the availability of high-quality forage in spring. But the vast majority of studies report a substantial overlap in the diet of introduced domesticates with native species (e.g. Mackie 1970, Hansen and Reid 1975, Hansen *et al.* 1977, Schwartz and Ellis 1981, Hanley and Hanley 1982, Ghosh *et al.* 1987) and where specific evidence has been sought of possible niche shifts or competitive suppression of population growth, results are generally indicative of some level of competition.

(Such overlap between native species and domestic stock is not a simple, universal phenomenon but is quite clearly structured: cattle and horses, as preferential grazers, tend to show highest overlap and thus potential for competition with native species reliant on a bulk-feeding strategy (*sensu* Hofmann 1973, 1985); competition for specialist browsers may be afforded by goats, while sheep are most likely to show high overlap with intermediate feeders.)

Thus while summer grazing by cattle may be seen to enhance forage quality for red deer hinds in the Isle of Rum, changes in patterns of resource use and partitioning of resources by red deer, cattle, ponies and goats between summer and winter are strongly suggestive of potential competition during the period of limited forage availability (Gordon and Illius 1989). A high degree of overlap in resource use was recorded between all the species over the summer period, when forage was abundant; over the winter period of low food availability far stricter partitioning of diets was observed, with different species exploiting relatively exclusive sets of resources. Diets of cattle introduced to dry–temperate pine–oak forests in Mexico showed little overlap with that of indigenous white-tailed deer, suggesting low potential for competition (Gallina 1984). By contrast, substantial overlap in forage use between cattle and white-tailed deer has been reported by others, and distinct and significant shifts in patterns of habitat use of mule deer were recorded by Loft *et al.* (1991) following the introduction of cattle to native rangelands in the Sierra Nevada mountains of California: with the size of niche shift directly related to intensity of cattle grazing. Mule deer reduced their selection of preferred habitats and increased their selection for habitats avoided by cattle as stocking densities increased.

All the various pieces of information we can piece together suggest that established natural communities are likely to be relatively free of competition. From a purely theoretical stand we may argue that natural selection would be expected to promote clear separation in resource use between regularly interacting sets of species specifically to minimize the loss of fitness incurred through competition. We would expect relatively little evidence of any interaction between species; those recorded are more likely to be facilitative than competitive. Competition should become apparent only when an established system is challenged by some perturbation of species composition or relative density. Even here such competition is likely not to exert any significant effect on population numbers or resource relationships of coexistent species if predation (or population control through human agency) is sufficient to suppress populations below levels at which critical resources may become limiting.

In the New Forest of Hampshire, there would appear to be little evidence for competitive interaction amongst the deer themselves. Yet this is not a natural assembly: both fallow deer and sika are introduced exotics. We must presume, as before, that lack of apparent competition amongst the 'wild' herbivores of the Forest is due to the fact that populations are maintained by artificial 'predation' to levels below those at which the particular resources for which they might compete amongst themselves become limiting. By contrast, populations of domestic stock are kept high by economic demand and artificial subsidy. In consequence resources used extensively by cattle and ponies (grasslands and grasses) do become limiting; where overlap in resource use between the Forest's ungulates is in relation to these particular commodities (as between cattle and ponies, or cattle and ponies with fallow, red or sika deer) explicit competition may become apparent.

References

Aitchison, J. (1982) Statistical analysis of compositional data. *Journal of the Royal Statistical Society B*, **44**, 139–77.

Andersen, J. (1953) Analysis of a roe deer population based on extermination of the total stock. *Danish Review of Game Biology*, **2**, 127–55.

Apollonio, M. (1989) Lekking in fallow deer: just a matter of density? *Ethology, Ecology and Evolution*, **1**, 291–4.

Apollonio, M., Festa-Bianchet, M. and Mari, F. (1989) Correlates of copulatory success in a fallow deer lek. *Behavioural Ecology and Sociobiology*, **25**, 89–92.

Apollonio, M., Festa-Bianchet, M., Mari, F. and Riva, M. (1990) Site-specific asymmetries in male copulatory success in a fallow deer lek. *Animal Behaviour*, **39**, 205–12.

Apollonio, M., Festa-Bianchet, M., Mari, F., Mattioli, S. and Sarno, B. (1992) To lek or not to lek: mating strategies of male fallow deer. *Behavioural Ecology*, **3**, 25–31.

Archer, M. (1973) Variations in potash levels in pastures grazed by horses; a preliminary communication. *Equine Veterinary Journal*, **5**, 45–6.

Atkinson, W.D. and Shorrocks, B. (1981) Competition on a divided and ephemeral resource: a simulation model. *Journal of Animal Ecology*, **50**, 461–71.

Atkinson, W.D. and Shorrocks, B. (1984) Aggregation of larval Diptera over discrete and ephemeral breeding sites: the implications for coexistence. *American Naturalist*, **124**, 336–51.

Bakker, J.P., de Bie, S., Dallinga, J.H., Tjaden, P. and de Vries, Y. (1983a) Sheep-grazing as a management tool for heathland conservation and regeneration in the Netherlands. *Journal of Applied Ecology*, **20**, 541–60.

Bakker, J.P., de Leeuw, J. and van Wieren, S.E. (1983b) Micro-patterns in grassland vegetation created and sustained by sheep grazing. *Vegetatio*, **55**, 153–61.

Bartos, L., Zeeb, U. and Mikes J. (1992) Lekking behaviour in sika deer, in *Wildlife Conservation: Present Trends and Perspectives for the 21st Century* (eds N. Maruyama *et al.*), Tokyo.

Bell, R.H.V. (1970) The use of the herb layer by grazing ungulates in the Serengeti, in *Animal Populations in Relation to Their Food Resources* (ed. A. Watson), Blackwell Scientific Publications, Oxford, pp. 111–23.

Ben-Shahar, R. and Skinner, J.D. (1988) Habitat preferences of African ungulates derived by uni- and multi-variate analyses. *Ecology*, **69**, 1479–85.

Berger, J. (1977) Organisational systems and dominance in feral horses in the Grand Canyon. *Behavioural Ecology and Sociobiology*, **2**, 131–46.

Berger, J. (1986) *Wild Horses of the Great Basin*, Chicago University Press.

Boxall, M. R. (1990) Patterns of habitat use and spatial separation of four deer species living in sympatry in Roydon Woods Nature Reserve, Hampshire. Dissertation submitted for Certificate in Field Biology, Birkbeck College, University of London.

Burton, D. (1992) The effects of parasitic nematode infection on body condition of New Forest ponies. PhD thesis, University of Southampton.

Cairns, A.L. and Telfer, E.S. (1980) Habitat use by 4 sympatric ungulates in boreal mixed wood forest. *Journal of Wildlife Management*, **44**, 849–57.

Calder, C. (1995) Habitat structure and roe deer foraging strategies. Can different diets be of equal nutritive value for roe deer? Presentation to 2nd European Congress of Mammalogy, Southampton March 1995.

Caldwell, J.F., Chapman, D.I. and Chapman, N.G. (1983) Observations on the autumn and winter diet of fallow deer (*Dama dama*). *Journal of Zoology (London)*, **201**, 559–63.

Challies, C.N. (1985) Establishment, control and commercial exploitation of wild deer in New Zealand, in *Biology of Deer Production* (eds K.R. Drew and P.F. Fennessy), Royal Society of New Zealand Bulletin, **22**, 23–36.

Chamrad, A.D. and Box, T.W. (1964) A point frame for sampling rumen content. *Journal of Wildlife Management*, **28**, 473–7.

Chapman, D.I. and Chapman, N.G. (1976) *Fallow Deer: Their History, Distribution and Biology*, Terence Dalton, Lavenham.

Chapman, D.I. and Chapman, N.G. (1980) The distribution of fallow deer: a worldwide review. *Mammal Review*, **10**, 61–138.

Chapman, N.G. and Putman, R.J. (1991) Fallow deer, in *The Handbook of British Mammals*, 3rd edn (eds G.B. Corbet and S. Harris), Blackwell, Oxford, pp. 508–18.

Chapman, N.G., Claydon, K., Claydon, M. and Harris, S. (1985) Distribution and habitat selection by muntjac and other species of deer in a coniferous forest. *Acta Theriologica*, **30**, 287–303.

Chapman, N.G., Claydon, K., Claydon, M., Forde, P.G. and Harris, S. (1993) Sympatric populations of muntjac and roe deer: a comparative analysis of their ranging behaviour, social organisation and activity. *Journal of Zoology*, **229**, 623–40.

Chesson, P.L. and Case, T.J. (1986) Non-equilibrium community theories: chance, variability, history and coexistence, in *Community Ecology* (eds J.M. Diamond and T.J. Case), Harper and Row, New York, pp. 229–39.

Clutton-Brock, T.H. and Albon, S.D. (1989) *Red Deer in the Highlands*, Blackwell Scientific Publications, Oxford.

Clutton-Brock, T.H., Deutsch, J.C. and Nefdt, R.J.C. (1993) The evolution of ungulate leks. *Animal Behaviour*, **46**, 1121–38.

Clutton-Brock, T.H., Guinness, F.E. and Albon, S.D. (1982) *Red Deer: Behaviour and Ecology of Two Sexes*, Edinburgh University Press/Chicago University Press.

Clutton-Brock, T.H., Hiraiwa-Hasegawa, M. and Robertson, A. (1989) Mate choice on fallow deer leks. *Nature (London)*, **340**, 463–5.

Clutton-Brock, T.H., Iason, G.R. and Guinness, F.E. (1987) Sexual segregation and density related changes in habitat use in male and female red deer. *Journal of Zoology (London)*, **211**, 275—89.

Clutton-Brock, T.H., Green, D., Hiraiwa-Hasegawa, M. and Albon, S.D. (1988) Passing the buck: resource defence, lek breeding and mate choice in fallow deer. *Behavioural Ecology and Sociobiology*, **23**, 281–96.

Connell, J.H. (1978) Diversity in tropical rain forests and coral reefs. *Science*, **199**, 1302–10.

Countryside Commission (1984) *The New Forest Commoners*, Countryside Commission, Cheltenham.

Dabrowski, M.J. (1959) Late-glacial and Holocene history of Bialowieza Prime Forest. *Acta Societatis Botanicorum Poloniae*, **28**, 197–248.

Danilkin, A. and Hewison, A.J.M. (1996) *Behavioural Ecology of Siberian and European Roe Deer*, Chapman & Hall Wildlife Ecology and Behaviour Series, Chapman & Hall, London.

de Boer, W.F. and Prins, H.H.T. (1990) Large herbivores that strive mightily but eat and drink as friends. *Oecologia*, **82**, 264–74.

Diakite, M.M. (1983) A study of the winter diet of roe deer (*Capreolus capreolus* L.) at Alice Holt Forest. M.Sc thesis, University College, London.

Duncan, P. (1983) Determinants of the use of habitat by horses in a Mediterranean wetland. *Journal of Animal Ecology*, **52**, 93–109.

Duncan, P. (1992) *Horses and Grasses; The Nutritional Ecology of Equids and Their Impact on the Camargue*, Springer-Verlag, New York and Berlin.

Dusek, G.L. (1975) Range relations of mule deer and cattle in prairie habitat. *Journal of Wildlife Management*, **39**, 605–16.

Edwards, P.J. and Hollis, S. (1982) The distribution of excreta on New Forest grasslands used by cattle, ponies and deer. *Journal of Applied Ecology*, **19**, 953–64.

Ekins, J.R. (1989) Forage resources of cattle and ponies in the New Forest, Southern England. PhD thesis, University of Southampton.

Ellis, R.N.W. (1975) Growth and nutrition studies in the pony. PhD thesis, University of Liverpool.

Farley, A.E. (1986) Diets of foxes in the south of England. B.Sc. Honours thesis (Biology), University of Southampton.

Feist, J.D. and McCullough, D.R. (1976) Behaviour patterns and communication in feral horses. *Zeitschrift für Tierpsychologie*, **41**, 337–71.

Feldhamer, G.A. and Armstrong, W.E. (1993) *Interspecific Competition Between Four Exotic Species and Native Artiodactyls in the United States*. Transactions of the 58th North American Wildlife and Natural Resources Conference, pp. 468–78.

Feldhamer, G.A., Chapman, J.A. and Miller, R.L. (1978) Sika and white-tailed deer on Maryland's eastern shore. *Wildlife Society Bulletin*, **6**, 155–7.

Field, C.R. (1968) A comparative study of the food habits of some wild ungulates in the Queen Elizabeth National Park, Uganda; preliminary report, in *Comparative Nutrition of Wild Animals* (ed. M.A. Crawford), Symposia of the Zoological Society of London, **21**, 135–51.

Field, C.R. (1972) The food habits of wild ungulates in Uganda by analysis of stomach contents. *East African Wildlife Journal*, **10**, 17–42.

Forde, P.G. (1989) The comparative ecology of muntjac (*Muntiacus reevesi*) and roe deer (*Capreolus capreolus*) in a commercial coniferous forest. PhD thesis, University of Bristol.

Fretwell, S.D. (1972) *Populations in a Seasonal Environment*, Princeton University Press, Princeton.

Fretwell, S.D. and Lucas, H.L. (1970) On territorial behaviour and other factors influencing habitat distribution in birds. *Acta Biotheoretica*, **19**, 16–36.

Furubayashi, K. and Maruyama, N. (1977) Food habits of sika deer in Fudakake, Tanzawa Mountains. *Journal of the Mammal Society of Japan*, **7**, 55–62.

Gallina, S. (1984) Ecological aspects of the coexploitation of deer *Odocoileus virginianus* and cattle (Proceedings of the 3rd International Theriological Congress, Helsinki, 1982). *Acta Zoologica Fennica*, **172**, 251–4.

Gates, S.A. (1980) Studies of the ecology of the free-ranging Exmoor Pony. PhD thesis, University of Exeter.

Gates, S.A. (1982) The Exmoor pony – a wild animal? *Nature in Devon (Journal of the Devon Trust for Nature Conservation)*, **2**, 7–30.

Gessaman, J.A. and MacMahon, J.A. (1984) Mammals in ecosystems: their effects on the composition and production of vegetation. *Acta Zoologica Fennica*, **172.**

Ghosh, P.K., Goyal, S.P. and Bohra, H.C. (1987) Competition for resource utilisation between wild and domestic ungulates in the Rajasthan desert. *Tigerpaper*, **14**, 2–7.

Gill, E.L. (1988) Factors affecting body condition of New Forest ponies. PhD thesis, University of Southampton.

Gill, E.L. (1991) Factors affecting body condition in free-ranging ponies. *Technical Report, Royal Society for the Prevention of Cruelty to Animals.*

Gordon, I.J. (1988) Facilitation of red deer grazing by cattle and its impact on red deer performance. *Journal of Applied Ecology*, **25**, 1–10.

Gordon, I.J. and Illius, A.W. (1989) Resource partitioning by ungulates on the Isle of Rhum. *Oecologia*, **79**, 383–90.

Grodzinski, W. (1975) The role of large herbivorous mammals in the functioning of forest ecosystems – a general model. *Polish Ecological Studies*, **1/2**, 5–15.

Gwynne, M.D and Bell, R.H.V. (1968) Selection of vegetation components by grazing ungulates in the Serengeti National Park. *Nature*, **220**, 390–3.

Hanley, T.A. and Hanley, K.A. (1982) Food resource partitioning by sympatric ungulates on Great Basin rangeland. *Journal of Range Management*, **35**, 152–8.

Hansen, R.M. (1976) Foods of free-roaming horses in southern New Mexico. *Journal of Range Management*, **29**, 347.

Hansen, R.M. and Reid, L.D. (1975) Diet overlap of deer, elk and cattle in southern Colorado. *Journal of Range Management*, **26**, 43–7.

Hansen, R.M., Clark, R.C. and Lawhorn, W. (1977) Foods of wild horses, deer and cattle in the Douglas Mountain area, Colorado. *Journal of Range Management*, **30**, 116–18.

Harmel, D.E. (1980) *The Influence of Exotic Artiodactyls on White-tailed Deer Performance and Survival*, Performance Report: Job No. 20, Federal Aid Project W-109-R-3, Texas Parks and Wildlife Department.

Harmel, D.E. (1992) *The Influence of Fallow Deer and Aoudad Sheep on White-tailed Deer Production and Performance*, Performance Report: Job No. 20, Federal Aid Project W-127-R-1, Texas Parks and Wildlife Department.

Harrington, R. (1982) The hybridisation of red deer (*Cervus elaphus* L. 1758) and Japanese sika deer (*C. nippon* Temminck, 1838). *International Congress of Game Biologists*, **14**, 559–71.

Harris, S., Morris, P., Wray, S. and Yalden D.W. (1995) *A Review of British Mammals: Population Estimates and Conservation Status of British Mammals other than Cetaceans*, Joint Nature Conservation Committee, Peterborough.

Heideman, G. (1973) *Zur Biologie des Damwildes*, Paul Parey, Hamburg and Berlin.

Hewison, A.J.M. (1996) Variation in the fecundity of roe deer: relative effects of habitat quality and climatic factors. *Acta Theriologica*, in press.

Hill, S.D. (1985) Influences of large herbivores on small rodents in the New Forest. PhD thesis, University of Southampton.

Hirons, G.J.M. (1984) The diet of tawny owls (*Strix aluco*) and kestrels (*Falco tinnunculus*) in the New Forest, Hampshire. *Proceedings of the Hampshire Field Club and Archaeological Society*, **40**, 21–6.

Hirst, S.M. (1975) Ungulate–habitat relationships in a South African woodland/savannah ecosystem. *Wildlife Monographs*, **44.**

Hirth, D.H. (1977) Social behaviour of white-tailed deer in relation to habitat. *Wildlife Monographs*, **53**, 1–55.

Hofmann, R.R. (1973) *The Ruminant Stomach,* East African Literature Bureau, Nairobi.

Hofmann, R.R. (1982) Morphological classification of sika deer within the comparative system of ruminant feeding types. *Deer,* **5**, 252–3.

Hofmann, R.R. (1985) Digestive physiology of the deer – their morphophysiological specialisation and adaptation, in *Biology of Deer Production* (eds K.R. Drew and P.F. Fennessy), Royal Society of New Zealand, Bulletin, **22**, 393–407.

Homolka, M. (1993) The food niches of three ungulate species in a woodland complex. *Folia Zoologica (Brno),* **42**, 193–203.

Horwood, M.T. and Masters, E.H. (1970) *Sika Deer,* British Deer Society, Reading.

Hosey, G.R. (1974) The food and feeding ecology of the roe deer (*Capreolus capreolus*). PhD thesis, University of Manchester.

Hosey, G.R. (1981) Annual foods of the roe deer (*Capreolus capreolus*) in the south of England. *Journal of Zoology (London),* **194**, 276–8.

Houston, D.B. (1982) *The Northern Yellowstone Elk: Ecology and Management,* Macmillan, New York.

Howard, P.C. (1979) Variability of feeding behaviour in New Forest ponies. B.Sc. Honours thesis (Biology), University of Southampton.

Hudson, R.J. (1976) Resource division within a community of large herbivores. *Le Naturaliste Canadien,* **103**, 153–67.

Hunter, R.F. (1962) Hill sheep and their pasture: a study of sheep grazing in south east Scotland. *Journal of Ecology,* **50**, 651–80.

Huston, M. (1979) A general hypothesis of species diversity. *American Naturalist,* **113**, 81–101.

Hutchinson, G.E. (1957) Concluding remarks. *Cold Spring Harbor Symposia in Quantitative Biology,* **22**, 415–27.

Illius, A.W. and Gordon, I.J. (1987) The allometry of food intake in grazing ruminants. *Journal of Animal Ecology,* **56**, 989–99.

Insley, H. (1977) An estimate of the population of the red fox (*Vulpes vulpes*) in the New Forest, Hampshire. *Journal of Zoology (London),* **183**, 549–53.

Jackson, J.E. (1974) The feeding ecology of fallow deer in the New Forest. PhD thesis, University of Southampton.

Jackson, J.E. (1977) The annual diet of the fallow deer (*Dama dama*) in the New Forest, Hampshire, as determined by rumen content analysis. *Journal of Zoology (London),* **181**, 465–73.

Jackson, J.E. (1980) The annual diet of the roe deer (*Capreolus capreolus*) in the New Forest, Hampshire, as determined by rumen content analysis. *Journal of Zoology (London),* **192**, 71–83.

Jarman, P.J. (1971) Diets of large mammals in the woodlands around Lake Kariba, Rhodesia. *Oecologia,* **8**, 157–78.

Jarman, P.J. (1972) Seasonal distribution of large mammal populations in the unflooded middle Zambesi valley. *Journal of Applied Ecology,* **9**, 283–99.

Jarman, P.J. (1974) The social organisation of antelope in relation to their ecology. *Behaviour,* **48**, 215–67.

Jarman, P.J. (1982) Prospects for interspecific comparison in sociobiology, in *Current Problems in Sociobiology* (ed. Kings College Cambridge Sociobiology Group), Cambridge University Press, Cambridge, pp. 323–42.

Jarman, P.J. and Jarman, M.V. (1979) The dynamics of ungulate social organization, in *Serengeti: Dynamics of an Ecosystem* (eds A.R.E. Sinclair and M. Norton-Griffiths), University of Chicago Press, Chicago, pp. 185–220.

Jarman, P.J. and Sinclair, A.R.E. (1979) Feeding strategy and the pattern of resource partitioning in ungulates, in *Serengeti: Dynamics of an Ecosystem* (eds A.R.E. Sinclair and M. Norton-Griffiths), University of Chicago Press, Chicago, pp. 130–63.

Jedrzejewska, B., Okarma, H., Jedrzejewski, W. and Milkowski, L. (1994) Effects of exploitation and protection on forest structure, ungulate density and wolf predation in Bialowieza Primeval Forest, Poland. *Journal of Applied Ecology*, **31**, 664–76.

Jedrzejewska, B., Jedrzejewski, W., Bunevich, A.N., Milkowski, L. and Krasinski, Z.A. (1996) Ungulates in Bialowieza primeval forest (Poland and Belarus) – 200 years of population dynamics. *Acta Theriologica*, in press.

Jedrzejewski, W., Jedrzejewska, B., Okarma, H. and Ruprecht, A.L. (1992) Wolf predation and snow cover as mortality factors in the ungulate community of the Bialowieza National Park, Poland. *Oecologia*, **90**, 27–36.

Jedrzejewski, W., Jedrzejewska, B., Okarma, H., Schmidt, K., Bunevich, A.N. and Milkowski, L. (1996) Population dynamics (1869–1994), demography and home ranges of the lynx in Bialowieza Primeval Forest (Poland and Belarus). *Ecography*, in press

Jenkins, K.J. and Wright, R.G. (1988) Resource partitioning and competition among cervids in the northern Rocky Mountains. *Journal of Applied Ecology*, **25**, 11-24.

Johnson, T.J. (1984) Habitat and social organisation of roe deer (*Capreolus capreolus*). PhD thesis, University of Southampton.

Kaluzinski, J. (1982) Dynamics and structure of a field roe deer population. *Acta Theriologica*, **27**, 385–408.

Kaseda, Y. (1983) Seasonal changes in time spent grazing and resting of Misaki horses. *Japanese Journal of Zootechnology and Science*, **54**, 464–9.

Kay, R.N.B. (1979) Seasonal changes of appetite in deer and sheep. *Agricultural Research Council Research Review 1979*.

Kay, R.N.B. and Staines, B.W. (1981) The nutrition of the red deer (*Cervus elaphus*). *Nutrition Abstracts and Reviews, Series B*, **51**, 601–21.

Kean, R.I. (1959) Ecology of the larger wildlife mammals of New Zealand. *New Zealand Science Reviews*, **17**, 35–7.

Keiper, R.R. (1985) Are sika deer responsible for the decline of white-tailed deer on Assateague Island, Maryland? *Wildlife Society Bulletin*, **13**, 144–6.

Kenchington, F.E. (1944) *The Commoner's New Forest*, Hutchinson, London.

Kiddie, D.G. (1962) *The Sika Deer* (Cervus nippon) *in New Zealand*, New Zealand Forest Information Series No. 44.

Kurt, F. (1978) *Das Sozialverhalten des Rehes* (Capreolus capreolus *L.*). Paul Parey, Hamburg and Berlin.

LaGory, K.E. (1986) Habitat, group size and the behaviour of white-tailed deer. *Behaviour*, **98**, 168–79.

Lamprey, H.F. (1963) Ecological separation of the large mammal species in the Tarangire Game Reserve, Tanganyika. *East African Wildlife Journal*, **1**, 63–92.

Langbein, J. (1991) Effects of density and age on body condition, reproductive performance, behaviour and survival of fallow deer. PhD thesis, University of Southampton.

Langbein, J. and Putman, R.J. (1992) Behavioural responses of park red and fallow deer to disturbance, and effects on population performance. *Animal Welfare*, **1**, 19–38.

Langbein, J. and Thirgood, S.J. (1989) Variation in mating systems of fallow deer in relation to ecology. *Ethology*, **83**, 195–214.

Langbein, J. and Thirgood, S.J. (in press) *Fallow Deer; Behavioural Plasticity in Contrasting Environments*, Chapman & Hall Wildlife Ecology and Behaviour Series, Chapman & Hall, London.

Lascelles, G.W. (1915) *Thirtyfive Years in the New Forest*, Arnold, London.

Leader-Williams, N. and Ricketts, C. (1982) Seasonal and sexual patterns of growth and condition of reindeer (*Rangifer tarandus*) introduced into South Georgia. *Oikos*, **38**, 27–39.

Leuthold, W. (1978) Ecological separation among browsing ungulates in the Tsavo East National Park, Kenya. *Oecologia*, **35**, 241–52.

Loft, E.R., Menke, J.W. and Kie, J.G. (1991) Habitat shifts by mule deer: the influence of cattle grazing. *Journal of Wildlife Management*, **55**, 16–25.

Lowe, V.P.W. and Gardiner, A.S. (1975) Hybridisation between red deer and sika deer, with reference to stocks in north-west England. *Journal of Zoology (London)*, **177**, 553–66.

MacArthur, R.H. and Levins, R. (1967) The limiting similarity, convergence and divergence of coexisting species. *American Naturalist*, **101**, 377–85.

McCullough, D.R. (1982) Evaluation of night spotlighting as a deer study technique. *Journal of Wildlife Management*, **46**, 963–73.

McCullough, D.R. (1985) Variables influencing food habits of white-tailed deer on the George Reserve. *Journal of Mammalogy*, **66**, 682–92.

McEwan, E.K. and Whitehead, P.E. (1970) Seasonal changes in the energy and nitrogen intake in reindeer and caribou. *Canadian Journal of Zoology*, **48**, 905–13.

McEwen, L.S., French, C.E., Magruder, N.D., Swift, R.W. and Ingram, R.H. (1957) *Nutrient Requirements of the White-tailed Deer*. Transactions of the 22nd North American Wildlife Conference, pp. 119-32.

McKelvey, P.J. (1959) Animal damage in North Island protection forests. *New Zealand Science Reviews*, **17**, 28–34.

Mackie, R.J. (1970) Range ecology and relations of mule deer, elk and cattle in the Missouri River Breaks, Montana. *Wildlife Monographs*, **20**, 1–79.

McNaughton, S. (1979) Grassland–herbivore dynamics, in *Serengeti: Dynamics of an Ecosystem* (eds A.R.E. Sinclair and M. Norton-Griffiths), Chicago University Press, pp. 46–81.

Mann, J.C.E. (1983) The social organisation and ecology of the Japanese sika deer (*Cervus nippon*) in Southern England. PhD thesis, University of Southampton.

Mann, J.C.E. and Putman, R.J. (1989a) Habitat use and activity patterns of British sika deer (*Cervus nippon* Temminck) in contrasting environments. *Acta Theriologica*, **34**, 83–96.

Mann, J.C.E. and Putman, R.J. (1989b) Diet of British sika deer (*Cervus nippon* Temminck) in contrasting environments. *Acta Theriologica*, **34**, 97–110.

Martin, C. (1987) Interspecific relationships between barasingha (*Cervus duvauceli*) and axis deer (*Axis axis*) in Kanha National Park, India, and relevance to management, in *Biology and Management of the Cervidae* (ed. C.M. Wemmer), Smithsonian, Washington, pp. 299–306.

Mathur, V.B. (1991) Ecological interaction between habitat composition, habitat quality and abundance of some wild ungulates in India. D.Phil. thesis, University of Oxford.

May, R.M. (1976/1981) *Theoretical Ecology*, Blackwell Scientific Publications.

Mayes, E. and Duncan, P. (1986) Temporal patterns of feeding behaviour in free-ranging horses. *Behaviour*, **96**, 105–29.

Miller, T.E. (1982) Community diversity and interactions between the size and frequency of disturbance. *American Naturalist*, **120**, 533–6.

Mishra, H.R. (1982) The ecology and behaviour of chital (*Axis axis*) in the Royal Chitawan National Park, Nepal. PhD thesis, University of Edinburgh.

Mitchell, B., McCowan, D. and Nicholson, I.A. (1976) Annual cycles of body weight and condition in Scottish red deer (*Cervus elaphus*). *Journal of Zoology (London)*, **180**, 107–27.

Moen, A.N. (1976) Energy conservation by white-tailed deer (*Odocoileus virginianus*) in the winter. *Ecology*, **57**, 192–8.

Moen, A.N. (1978) Seasonal changes in heart rate, activity, metabolism and forage intake of white-tailed deer (*Odocoileus virginianus*). *Journal of Wildlife Management*, **42**, 715–38.

Okarma, H., Jedrzejewska, B., Jedrzejewski, W., Krasinski, Z.A. and Milkowski, L. (1995) The roles of predation, snow cover, acorn crop and man-related factors on ungulate mortality in Bialowieza Primeval Forest, Poland. *Acta Theriologica*, **40**, 197–217.

Payne, D.R. (1987) Aspects of the social behaviour and ecology of the New Forest red deer. Dissertation submitted for University Certificate in Environmental Studies, University of Southampton.

Peterken, G.F. and Tubbs, C.R. (1965) Woodland regeneration in the New Forest, Hampshire, since 1650. *Journal of Applied Ecology*, **2**, 159–70.

Petrusewicz, K. (ed.) (1967) *Secondary Productivity in Terrestrial Ecosystems*, Panstwowe Wydawnictwo Naukowe, Warsaw.

Petty, S.J. and Avery, M.I. (1990) *Forest Bird Communities*, Occasional Paper 26, Forestry Commission, Edinburgh.

Pianka, E.R. (1973) The structure of lizard communities. *Annual Review of Ecology and Systematics*, **4**, 53–74.

Pianka, E.R. (1976) Competition and niche theory, in *Theoretical Ecology* (ed. R.M. May), Blackwell Scientific Publications, pp. 114–41.

Pianka, E.R. (1981) Competition and niche theory, in *Theoretical Ecology*, 2nd edn (ed. R.M. May), Blackwell Scientific Publications, pp. 167–96.

Pickering, D.W. (1968) Heathland reclamation in the New Forest: the ecological consequences. M.Sc. thesis, University College, London.

Pollock, J.I. (1980) Behavioural ecology and body condition changes in New Forest ponies. *RSPCA Scientific Publications No. 6*, Royal Society for the Prevention of Cruelty to Animals, Horsham.

Pratt, R.M., Putman, R.J., Ekins, J.R. and Edwards, P.J. (1986) Habitat use of free-ranging cattle and ponies in the New Forest of southern England. *Journal of Applied Ecology*, **23**, 530–57.

Prior, R. (1968) *The Roe Deer of Cranborne Chase*, Oxford University Press, Oxford.

Prior, R. (1973) Roe deer management and stalking. *Game Conservancy Booklet 17*, Game Conservancy Trust, Fordingbridge.

Prisyazhynuk, V.E. and Prisyazhynuk, N.P. (1974) [Sika deer on Askold Island]. *Bulletin Moskow o-va ispyt. Priv. otd. Biology*, **79**, 16–27 (in Russian).

Putman, R.J. (1981) Social systems of deer: a speculative review. *Deer*, **5**, 186–8.

Putman, R.J. (1984) Facts from faeces. *Mammal Review*, **14**, 79–97.

Putman, R.J. (1986a) Competition and coexistence in a multispecies grazing system. *Acta Theriologica*, **31**, 271–91.

Putman, R.J. (1986b) *Grazing in Temperate Ecosystems; Large Herbivores and their Effects on the Ecology of the New Forest*, Croom Helm/Chapman & Hall, London.

Putman, R.J. (1987) The impact of grazing by large herbivores: a case study in the New Forest of southern England, in *Begrazing in der Natuur* (eds S. de Bie, W. Joenje and S.E. van Wieren), Pudoc, The Netherlands, pp. 187–99.

Putman, R.J. (1988) *The Natural History of Deer*, Christopher Helm/Academic Press, London.
Putman, R.J. (1991) Horses, in *The Handbook of British Mammals*, 3rd edn (eds G.B. Corbet and S. Harris), Blackwell, Oxford, pp. 486–91.
Putman, R.J. (1993) Flexibility of social organisation and reproductive strategy in deer. *Deer*, **9**, 23–8.
Putman, R.J. (1994) *Community Ecology*, Chapman & Hall, London.
Putman, R.J. and Hunt, E. (1994) Hybridisation between red and sika deer in Britain. *Deer*, **9**, 104–10.
Putman, R.J. and Mann, J.C.E. (1990) Social organisation and behaviour of British sika deer in contrasting environments. *Deer*, **8**, 90–4.
Putman, R.J. and Sharma, S.K. (1987) Long term changes in New Forest deer populations and correlated environmental change, in *Mammal Population Studies* (ed. S. Harris), Symposia of the Zoological Society of London, **58**, 167–79.
Putman, R.J., Culpin, S. and Thirgood, S.J. (1993) Sexual differences in composition and quality of the diets of male and female fallow deer in sympatry and in allopatry. *Journal of Zoology (London)*, **229**, 267–75.
Putman, R.J., Edwards, P.J., Pratt, R.M. and Ekins, J.R. (1981) Interrelationships between large herbivores of the New Forest and the Forest vegetation. *Report HF3/03/127 to Chief Scientists Team, Nature Conservancy Council.*
Putman, R.J., Langbein, J., Hewison, A.J.M. and Sharma, S.K. (1996) Relative roles of density-dependent and density-independent factors in population dynamics of British deer. *Mammal Review*, **26**, 81–101.
Putman, R.J., Pratt, R.M., Ekins, J.R. and Edwards, P.J. (1984) Patterns of habitat use and grazing by cattle and ponies and impact upon vegetation (Proceedings of the 3rd International Theriological Congress, Helsinki, 1982). *Acta Zoologica Fennica*, **172**, 183–6.
Putman, R.J., Pratt, R.M., Ekins, J.R. and Edwards, P.J. (1987) Food and feeding behaviour of cattle and ponies in the New Forest, Hampshire. *Journal of Applied Ecology*, **24**, 369–80.
Putman, R.J., Edwards, P.J., Mann, J.C.E., How, R.C. and Hill, S.D. (1989) Vegetational and faunal changes in an area of heavily grazed woodland following relief of grazing. *Biological Conservation*, **47**, 13–32.
Quirke, K. (1991) The diet of red deer, sika deer and Scottish blackface sheep in Killarney National Park, Co. Kerry. M.Sc. thesis, National University of Ireland.
Ratcliffe, P.R. (1987a) Distribution and current status of sika deer, *Cervus nippon*, in Great Britain. *Mammal Review*, **17**, 39–58.
Ratcliffe, P.R. (1987b) *The Management of Red Deer in Upland Forests*, Forestry Commission Bulletin No. 71, HMSO.
Ratcliffe, P.R. (1989) The control of red and sika deer populations in commercial forests, in *Mammals as Pests* (ed. R.J. Putman), Chapman & Hall, London, pp. 98–115.
Ratcliffe, P.R. and Mayle, B.A. (1992) *Roe Deer Biology and Management*, Forestry Commission Bulletin 105, HMSO.
Ratcliffe, P.R., Hall, J. and Allen, J. (1986) Computer predictions of sequential growth changes in commercial forests as an aid to wildlife management, with reference to red deer. *Scottish Forestry*, **40**, 79–83.
Ratcliffe, P.R., Peace, A.J., Hewison, A.J.M., Hunt, E.J. and Chadwick, A.H. (1992) The origins and characterization of Japanese sika deer populations of Great Britain, in *Wildlife Conservation: Present Trends and Perspectives for the 21st Century* (ed. N. Maruyama *et al.*), Tokyo, pp. 185–90.

Reynoldson, T.B. and Bellamy, L.S. (1970) The establishment of interspecific competition in field populations, with an example of competition in action between *Polycelis nigra* (Mull.) and *Polycelis tenuis* (Ijima), in *Dynamics of Populations* (eds P.J. den Boer and G.R. Gradwell), Pudoc, Wageningen, pp. 282–90.

Rhodes, B.D. and Sharrow, S.H. (1990) Effect of grazing by sheep on the quantity and quality of forage available to big game in Oregon coast range. *Journal of Range Management*, **43**, 235–7.

Rutberg, A.T. (1990) Inter-group transfer in Assateague pony mares. *Animal Behaviour*, **40**, 945–52.

Salt, G.W. (ed.) (1984) *Ecology and Evolutionary Biology: A Round Table on Research*, Chicago University Press.

Salter, R.E. and Hudson, R.J. (1978) Habitat utilisation by feral horses in western Alberta. *Le Naturaliste Canadien*, **105**, 309–21.

Salter, R.E. and Hudson, R.J. (1979) Feeding ecology of feral horses in western Alberta. *Journal of Range Management*, **32**, 221–5.

Schaaf, D. (1978) Some aspects of the ecology of the swamp deer or barasingha (*Cervus d. duvauceli*) in Nepal, in *Threatened Deer*, IUCN, Switzerland, pp. 66–86.

Schaal, A. (1982) Influence de l'environnement sur les composantes du groupe social chez le daim, *Cervus (Dama) dama* L. *Revue d'Ecologie (La Terre et la Vie)*, **36**, 161–74.

Schaal, A. (1986) Mise en évidence d'un comportement de reproduction en arène chez le Daim d'Europe (*Dama d. dama*). *Comptes Rendus de l'Academie des Sciences, Paris, Series III*, **18**, 729–32.

Schaal, A. and Bradbury J.W. (1987) Lek-breeding in a deer species. *Biology of Behaviour*, **12**, 28–31.

Schaller, G.B. (1972) *The Serengeti Lion: a Study of Predator–Prey Relations*, Chicago University Press Wildlife Behavior and Ecology Series, Chicago University Press.

Schmidt, P.J. (1969) Observations on the behaviour of cattle in a hot dry region of the Northern Territory of Australia, with particular reference to walking, watering and grazing. M.Sc. thesis, University of New England, Armidale, New South Wales.

Schoener, T.W. (1983) Field experiments on interspecific competition. *American Naturalist*, **122**, 240–85.

Schwartz, C.C. and Ellis, J.E. (1981) Feeding ecology and niche separation in some native and domestic ungulates on the shortgrass prairie. *Journal of Applied Ecology*, **18**, 343–54.

Shank, C.C. (1982) Age–sex differences in the diets of wintering Rocky Mountain bighorn sheep. *Ecology*, **63**, 627–33.

Sharma, S.K. (1994) The decline of the roe deer (*Capreolus capreolus* L.) in the New Forest Hampshire. PhD thesis, University of Southampton.

Shepperd, J.H. (1921) The trail of the short grass steer. *North Dakota Agricultural College Bulletin*, **154**, 8.

Short, H.L., Newsom, J.D., McCoy, G.L. and Fowler, J.F. (1969) *Effects of Nutrition and Climate on Southern Deer*. Transactions of the 34th North American Wildlife Conference, 137–45.

Sinclair, A.R.E. (1977) *The African Buffalo*, University of Chicago Press Wildlife Behavior and Ecology Series, University of Chicago.

Sinclair, A.R.E. (1979) Dynamics of the Serengeti Ecosystem, in *Serengeti: Dynamics of an Ecosystem* (eds A.R.E. Sinclair and M. Norton-Griffiths), University of Chicago Press, Chicago, pp. 1–30.

Sinclair, A.R.E. (1985) Does interspecific competition or predation shape the African ungulate community? *Journal of Animal Ecology*, **54**, 899–918.

Sinclair, A.R.E. and Norton-Griffiths, M. (1982) Does competition or facilitation regulate migrant ungulate populations in the Serengeti? A test of hypotheses. *Oecologia*, **53**, 364–69.

Sokal, R.R. and Rohlf, F.J. (1981) *Biometry*, 2nd edn, W.H. Freeman, New York.

Southwood, T.R.E., Brown, V.K. and Reader, P.M. (1979) The relationships of plant and insect diversities in succession. *Biological Journal of the Linnean Society*, **12**, 327–48.

Spedding, C.R.W. (1971) *Grassland Ecology*, Clarendon Press, Oxford.

Spiller, D.A. (1984) Seasonal reversal of competitive advantage between two spider species. *Oecologia*, **64**, 322–31.

Staines, B.W. and Crisp, J.M. (1978) Observations on food quality in Scottish red deer (*Cervus elaphus*) as determined by chemical analysis of the rumen contents. *Journal of Zoology (London)*, **185**, 253–9.

Staines, B.W. and Ratcliffe, P.R. (1987) Estimating the abundance of red deer (*Cervus elaphus* L.) and roe deer (*Capreolus capreolus* L.) and their current status in Great Britain, in *Mammal Population Studies* (ed. S. Harris), Symposium of the Zoological Society of London, **58**, 131–52.

Staines, B.W., Crisp, J.M. and Parish, T. (1982) Differences in the quality of food eaten by red deer (*Cervus elaphus*) stags and hinds in winter. *Journal of Applied Ecology*, **19**, 65—77.

Stelfox, J.G., Peden, D.G., Epp, H., Hudson, R.J., Mbugua, S.W., Agatsiva, J.L. and Amuyunzu, C.L. (1985) Herbivore dynamics in southern Narok, Kenya. *Journal of Wildlife Management*, **50**, 339–47.

Stewart, D.R.M. (1967) Analysis of plant epidermis in faeces: a technique for studying the food preferences of grazing herbivores. *Journal of Applied Ecology*, **4**, 83–111.

Stewart, D.R.M. and Stewart, J. (1970) Food preference data by faecal analysis for African plains ungulates. *Zoologica Africana*, **5**, 115.

Storr, G.M. (1961) Microscopic analysis of faeces: a technique for ascertaining the diet of herbivorous mammals. *Australian Journal of Biological Science*, **14**, 157–64.

Strange, M.L. (1976) A computer study of the fallow deer of the New Forest. B.Sc. Honours thesis (Environmental Sciences), University of Southampton.

Strong, D.R., Simberloff, D., Abele, L.G. and Thistle, A.B. (eds) (1984) *Ecological Communities: Conceptual Issues and the Evidence*. Princeton University Press, Princeton.

Suttie, J.M. and Simpson, A.M. (1985) Photoperiodic control of appetite, growth antlers and endocrine status of red deer (*Cervus elaphus*), in *Biology of Deer Production* (eds K.R. Drew and P.F. Fennessy), Royal Society of New Zealand Bulletin, **22**, 429–32.

Takatsuki, S. (1980) Food habits of sika deer on Kinkazan Island. *Scientific Reports of Tohoku University, 4th series*, **38**, 7–31.

Takatsuki, S. (1987) The general status of sika deer in Japan. *Deer*, **7**, 70—2.

Thirgood, S.J. (1990) Variation in social systems of fallow deer. PhD thesis, University of Southampton.

Thirgood, S.J. (1991) Alternative mating strategies and reproductive success in fallow deer. *Behaviour*, **116**, 1–10.

Thirgood, S.J. (1995a) The effects of sex, season and habitat availability on patterns of habitat use by fallow deer. *Journal of Zoology (London)*, **235**, 645–59.

Thirgood, S.J. (1995b) Ecological factors influencing sexual segregation and group size in fallow deer (*Dama dama*). *Journal of Zoology (London)*, in press.

Tubbs, C.R. (1968) *The New Forest: An Ecological History*, David and Charles, Newton Abbot.

Tubbs, C.R. (1974) *The Buzzard*, David and Charles, Newton Abbot.

Tubbs, C.R. (1982) The New Forest: conflict and symbiosis. *New Scientist*, **1 July**, 10–13.

Tubbs, C.R. (1986) *The New Forest: A Natural History*, Collins, London.

Tubbs, C.R. and Tubbs, J.M. (1985) Buzzards (*Buteo buteo*) and land use in the New Forest, Hampshire, England. *Biological Conservation*, **31**, 46–65.

Turner, D.C. (1979) An analysis of time-budgetting by roe deer (*Capreolus capreolus*) in an agricultural area. *Behaviour*, **71**, 246–90.

Tyler, S. (1972) The behaviour and social organisation of the New Forest ponies. *Animal Behaviour Monographs*, **5**, 87–194.

Vandermeer, J.H. (1972) Niche theory. *Annual Review of Ecology and Systematics*, **3**, 107–32.

van de Veen, H.E. (1979) Food selection and habitat use in the red deer (*Cervus elaphus* L.). PhD thesis, Rijksuniversiteit te Groningen, The Netherlands.

Wahlstrom, K. (1995) Dispersion and habitat use: an evolutionary perspective. *Forest Ecology and Management*, in press

Walther, F.R. (1972) Social grouping in Grant's gazelle (*Gazella granti* Brooke 1827) in the Serengeti National Park. *Zeitschrift für Tierpsychologie*, **31**, 348–403.

Waterfield, M.R. (1986) Observations on the ecology and behaviour of fallow deer (*Dama dama* L.). PhD thesis, University of Exeter.

Wells, S. and von Goldschmidt-Rothschild, B. (1979) Social behaviour and relationships in a group of Camargue horses. *Zeitschrift für Tierpsychologie*, **49**, 363–80.

Welsh, D. (1975) Population, behavioural and grazing ecology of the horses of Sable Island, Nova Scotia. PhD thesis, Dalhousie University.

Wiens, J.A. (1989) *The Ecology of Bird Communities. Volume 2: Processes and variations*. Cambridge University Press, Cambridge.

Wilkinson, G.S. (1986) Social grooming in the common vampire bat, *Desmodus rotundus*. *Animal Behaviour*, **34**, 1880–9.

Wood, A.J., Cowan, I.McT. and Nordan, H.C. (1962) Periodicity of growth in ungulates as shown by deer of the genus *Odocoileus*. *Canadian Journal of Zoology*, **40**, 596–603.

Wray, S. (1992) The ecology and management of European hares (*Lepus europaeus*) in commercial coniferous forestry. PhD thesis, University of Bristol.

Wray, S. (1994) Competition between muntjac and other herbivores in a commercial coniferous forest. *Deer*, **9**, 237–42.

Zejda, J. (1978) Field grouping of roe deer (*Capreolus capreolus*) in a lowland region. *Folia Zoologica (Brno)*, **27**, 111–22.

Index